Olga Kharaishvili

Corn Crop Forecasting

Olga Kharaishvili

Corn Crop Forecasting

LAP LAMBERT Academic Publishing

Imprint
Any brand names and product names mentioned in this book are subject to trademark, brand or patent protection and are trademarks or registered trademarks of their respective holders. The use of brand names, product names, common names, trade names, product descriptions etc. even without a particular marking in this work is in no way to be construed to mean that such names may be regarded as unrestricted in respect of trademark and brand protection legislation and could thus be used by anyone.

Cover image: www.ingimage.com

Publisher:
LAP LAMBERT Academic Publishing
is a trademark of
Dodo Books Indian Ocean Ltd. and OmniScriptum S.R.L publishing group

120 High Road, East Finchley, London, N2 9ED, United Kingdom
Str. Armeneasca 28/1, office 1, Chisinau MD-2012, Republic of Moldova, Europe
Managing Directors: Ieva Konstantinova, Victoria Ursu
info@omniscriptum.com

Printed at: see last page
ISBN: 978-620-9-08701-1

From the author

The predicted yield of corn depends on the variety or hybrid type, plant water supply, soil fertility, humus content, mineral composition and the use of modern technologies.

At the modern stage of development of agro-technological production, the stable predicted growth of agricultural crops, including corn, is ensured by the use of intensive technologies and maximum consideration of factors affecting the yield. The present monograph discusses:

- The current state of the predicted yield of corn;
- The agro-melioration characteristics of the Mukhran-Saguramo valley have been studied;
- A plan for conducting stationary field experiments has been developed;
- A study of predictive yield has been conducted;
- The methodology of the applied agrotechnical measures has been discussed;
- The economic efficiency of corn cultivation has been determined;
- The economic efficiency of corn cultivation with a predicted yield has been substantiated.

The monograph will be of great assistance to farmers employed in the agricultural sector, scientific specialists and persons interested in corn cultivation. The work has been approved and discussed at the Academic Council of Tbilisi Teaching University under the protocol N#6 on 15.01.2019 and is recommended for bachelors, masters and doctoral students of agricultural engineering, forestry, agronomy.

In order to further improve the work, the author will be grateful for the business comments, opinions and suggestions expressed.

Contents

Introduction

The prospects for Georgia's economic and social development are closely linked to the intensification of agricultural production and, first of all, to increasing the yield of grain crops.

Among grain crops, corn ranks third in the world in terms of area occupied, after wheat and rice. It is distinguished by its high yield, versatile food and nutritional values, and wide possibilities of use in many industries as a food product and raw material. Corn is characterized by great area flexibility, it can be grown in a variety of sharply different soil and climatic conditions. Corn grains contain a large amount of such valuable elements as carbohydrates, proteins, fats, as well as vitamins, etc.

The above-mentioned merits should explain the great interest that many generations of prominent scientists-researchers and practicing specialists have associated with the complex study of the properties of corn culture and the factors determining its yield.

In Georgia, corn is a universal crop, it is of decisive importance in grain production and in increasing livestock productivity.

Corn is the most productive compared to other grain crops. It uses a high agrotechnical background especially well in conditions of optimal management of the soil water regime.

Chapter I.
Current status of corn yield

1.1 Factors determining corn yield

The forecasted yield is based on a differentiated, simultaneous assessment of the mutually determining yield-producing multivariate factor using mathematical models and electronic computing machines, which, to some extent, ensures the transition from experimental agronomy to theoretical agronomy using personal computers.

Therefore, it is necessary to process the initial information accumulated in various fields of related sciences to determine the corresponding characteristics of the factors determining the process, on the basis of which an orderly system of complex measures is formed to obtain both real and maximum possible guaranteed harvest with a reliable probability under given specific soil and climatic conditions.

The modern approach to managing the process of growing agricultural crops involves quantitative assessment of the impact of natural and artificial factors on yield, complex accounting of agro-technological parameters, operational control of production processes, etc.

According to a predictable, i.e. pre-compiled program, which reflected the variation of the main factors for obtaining a crop, the first experiments were carried out in the 30s of the last century. A cropping system was developed, which, under typical soil and climatic conditions, ensured a corn grain yield of 10-12 tons per hectare.

Based on the data of long-term field observations, a corn biomass growth schedule was compiled, in accordance with which the plant's nutrition, water supply, water-air and carbon-oxygen exchange regime was regulated. The corn cultivation program fully reflected the dynamics of the formation of biological properties in individual phases of its growth and development.

In the experimental fields of many agricultural exhibitions, a corn yield of 10.0 t/ha was demonstrated, based on a pre-compiled crop structural formula that took into account the plant stand frequency, the number of productive stalks, cobs and grains placed in it. It also determined the average statistical mass of a group of 1000 grains in accordance with the plant water-fertilization schedule and the optimal doses of mineral fertilizers required to obtain a given crop. These studies should be considered the beginning of predictable crop yields. In this direction, large-scale scientific research work has been carried out in large scientific research centers in recent years, which was significantly facilitated by the development of computer technology and the creation of appropriate programs.

Initially, the basis for crop programming in specific bioclimatic conditions was various monofactors, which integrally reflected the influence of other factors on crop yield.

One of the first studies, taking into account the optimal photosynthetic potential of crops for programmed crop yield, was carried out at the end of the last century [68, 83] and was reflected in appropriate calculations and recommendations.

Later, these studies were expanded by introducing balance calculations of optimal nutrition and water use regimes in various works [30, 31]. In parallel, the development of methods for programmed crop yield began with a differentiated quantitative assessment and consideration of the characteristics of the main determinants of crop yield.

A large number of published monographs and scientific works are devoted to the theory of energy and mass variability in the soil-plant-atmosphere system, and to the issues of modeling the formation of plant productivity taking into account physiologically active radiation.

Various works have established agrophysical, agrometeorological, agrotechnical and other foundations for predicting corn yields using technological process management systems in agriculture, crop production and field food production.

It is emphasized in the monographs of Georgian and foreign scientists that crop programming serves to improve the organization of agrophytocenosis as a system for achieving its maximum productivity and includes the following issues:

- Preliminary calculation of crop formation and productivity, taking into account the physical-geographical, soil-climatic conditions of a given region, the biopotential and specific properties of certain plant/crop varieties, obtaining a high-quality crop with low self-costs with minimal labor, material-technical, energy and other resources;
- Application of the multifactor field-natural expert planning method in order to obtain objective information about the regularities of interaction of the main factors that form the crop;
- Development of agro-complex measures for growing agricultural crops in field and food crop rotation and compilation of network schedules based on optimization of technological processes of work.

  Programming, in addition to forecasting the quantity and quality of the crop, in advance, dynamically determines the state of its formation, that is, the organogenesis of plant growth and development according to stages and phases.

Based on the study and generalization of long-term and versatile experimental materials on the use of active radiation (PAR) for photosynthesis, scientists have formulated the following basic principles of crop programming:

- Determination of hydrothermal indices, which are a synthesis of two meteorological factors and complex characteristics of humidity. Studies have shown that the yield of many biological masses on Earth varies within a wide range. In particular, in tropical-humid regions it is equal to 50.0 t/ha, and in arid zones it is equal to 40.0 t/ha.

In terms of the use of photosynthetically active radiation (PAR), crops are divided into three groups:^ ordinary, which use 0.5-1.5% of it, good – 1.5-3.0%, record – 3.5-5.0% and theoretically possible – 8-10%. The higher the PAR utilization coefficient, the higher the biomass yield.

One of the reasons for the low values of the PAR utilization coefficient can be considered low humidity or its lack, nutrient deficiency, and the use of agricultural machinery for the crop or variety. It is also influenced by the geographical latitude and altitude of the area.

For maximum use of solar energy in agricultural production, it is necessary to form crops that will have the ability to accumulate more PAR.

Determination of the crop size according to the water supply of crops, which implies the establishment of the optimal irrigation regime in irrigated agriculture, and in non-irrigated conditions, determining the yield level, based on the established climatic conditions.

The determination of the crop size according to the PFR depends on and is closely related to the net productivity of photosynthesis (NPP), which fully characterizes the productivity of the ``work" of leaves. Studies have shown that in the temperate zone of grain production, every 1000 units of photosynthetic potential provide the formation of 3.5-3.0 kg of grain.

The crop size can be determined by the potential of agrophotosynthesis and the composition of crops on stubble and mowed fields. The yield is determined not only by the biological characteristics of the crop (specific variety), but also by the characteristics of care and cultivation, therefore this principle is based on the following laws of agriculture:

- Development of a system of fertilization for effective soil fertility, taking into account the demand for nutrients;
- Development of a complex of agrotechnical measures that ensures the receipt

of a programmed crop;

- Complete accounting of the main regularities of agriculture and plant breeding and conducting a preliminary analysis;
- Development of specific measures for the prevention and control of plant diseases and pests;
- Determination of optimal options for agrotechnical complexes necessary for obtaining a programmed crop;

Based on the above principles, the physiological, biological, agrochemical, agrophysical, agro-improvement and agrotechnical foundations of crop programming were developed, which have found wide application in agronomy and related fields, in particular in plant breeding and food production.

The physiological characteristics of crop programming provide for the formation of crops with optimal leaf area, net photosynthetic productivity (NPP) and photosynthetic potential of the assimilation surface (PP).

A high yield of corn grain can be achieved when the seedlings are capable of producing 3.0-3.5 kg of grain per 1000 m^2/ha per day. Such seedlings should have a maximum leaf area (50-55 thousand m2/ha) during the tillering phase.

According to literary sources, in non-irrigated fields, the area of corn leaves can be 13.1 thousand m2/ha, the productivity of photosynthesis per day - 4.2 g/m2, and the dry mass yield - 2.8 t/ha. In the case of irrigation and the application of 8,000 kg of humus and 90 kg of phosphorus fertilizer per hectare, the maximum leaf area was 32.7 thousand m2/ha, the productivity of photosynthesis - 8.5 g/m and the dry biomass - 12.24 t/ha. A further increase in corn leaf area to 62,000 m2/ha under irrigated conditions resulted in a decrease in photosynthetic productivity to 6.1 g/m2/day, but due to the large leaf area, a yield of 3.08 t/ha of dry biomass and 10.0 t/ha of corn grain was obtained. The following example shows the need to design a crop with a frequency that will maintain an optimal leaf area with a high photosynthetic efficiency (M.E.C.) over a long period of time.

The biological basis of crop programming involves the optimization of soil water-air, thermal and nutritional regimes, the management of environmental factors of plant habitats and the realization of the potential productivity of modern varieties and hybrids of agricultural crops.

Currently, crop production is subject to control and regulation. The works of Shalva Chanishvili and Aleksandre Ivanov and others clearly indicate that under natural conditions, yield programming over entire vast crop areas should be carried out by regulating individual factors or groups of factors (based on the condition of irreplaceability of factors), which ensures high yields of agricultural crops, including corn.

Agrochemical foundations of corn crop programming involve the justification of economically advantageous optimal doses of fertilizers for obtaining high yields, taking into account soil chemical indicators, crop nutrient uptake, as well as the coefficient of utilization of nutrients from the soil and applied fertilizers, soil fertility enhancement, and environmental protection requirements.

A number of other works have revealed the existence of a correlation between yield and output elements.

World agricultural practice shows that without the use of fertilizers, in the case of intensive cultivation, soil fertility rapidly decreases. Therefore, it is necessary to calculate fertilizer doses not only for obtaining a quick harvest, but also taking into account the preservation of soil fertility.

Agrophysical foundations of crop programming take into account the improvement of soil water-physical properties, bulk density, porosity, water capacity, heat capacity and other indicators. Studies have shown that soil compaction negatively affects plant growth and development.

Soil structuring is accompanied by a decrease in bulk density, improvement of soil water and physical properties, increased water permeability and, accordingly, the possibility of atmospheric precipitation and irrigation water entering the soil body, which contributes to the accumulation of moisture, a decrease in the share of strongly bound water (maximum molecular water capacity and hygroscopicity), an increase in the productive water supply in the active layer, a decrease in water evaporation from the soil surface, etc.

Weakly humified soils are characterized by low heat capacity and increased thermal conductivity. This, of course, makes it difficult to accumulate heat in the soil necessary for normal plant growth and development, which is especially necessary in cold and insufficiently warm periods of the year. Enriching the soil with humus contributes to the creation of a dense, water-resistant tower structure, which, of course, radically improves its important properties.

The agrometeorological basis of crop programming involves identifying the potential capabilities of climatic conditions, determining their characteristic qualitative and quantitative indicators (radiation balance, sum of active temperatures, atmospheric precipitation, etc.). Knowledge of these parameters is necessary for predicting the total productivity of crops, as well as weather conditions during the growing season (frost, drought, etc.).

According to the recommendations of scientists, the sum of active temperatures accumulated during the growing season above 10°C should be used for crop programming. In addition, the so-called bioclimatic potential (BCP) is used to assess the biological productivity of the climate, which, in addition to the sum of temperatures, takes into account the climatic layer and moisture supply.

Along with the forecast of the water regime for a given year, the forecast of physiologically active radiation (PAR) plays a major role in achieving maximum use of the climatic potential of the regions of Georgia.

Studies have shown that under conditions of high physiologically active radiation, moisture depletion causes droughts, which, of course, is associated with large economic losses due to the reduction of biomass.

Various methods of operational weather forecasting are widely covered in the works of Georgian and foreign scientists. In addition, they also note that all recommendations for forecasting meteorological conditions are predominantly local in nature and require adaptation to specific conditions.

The agrotechnical basis of crop programming provides for the timely and high-quality implementation of the entire agrotechnical complex of works, which should ensure the maximum disclosure of the plant's bio-potential through the use of optimal crop cultivation technologies.

In addition, important factors in the crop-producing process are: sowing norms, plant establishment frequency, sowing dates and rules, irrigation or drying regime, etc., which are subject to regulation, which gives the agrophytocenosis the character of a regulated and manageable object. One of the goals of programming is to form crops that provide high productivity with minimal costs.

Crop programming begins with the justification of obtaining a possible harvest. It is known that the harvest is formed in the process of photosynthesis. The level of yield depends on the biological properties of the culture (specific variety), the size of the canopy, the amount of moisture and nutrients in the soil, the level of agricultural technology and meteorological factors.

When planning the harvest of any agricultural crop, including corn, three levels of yield are usually determined:

- Potential yield (PY), obtained through synthetically active radiation;
- Actual yield (AY), using bioclimatic indicators and water supply parameters;
- Yield in production (YP). The level of yield obtained in production.

Potential yield (PY) is the theoretically possible maximum yield that can be obtained under ideal meteorological conditions (water, light, heat supply).

Actual yield (AY) is the maximum yield that can be obtained under real, average multi-year climatic conditions. Based on the analysis of many years of observations, it was determined that the MNP is 60-80% of the PM. The determination of the MNP is called crop programming.

Crop programming is worked out in detail, taking into account the agrochemical indicators of the soil, mineral and organic fertilizer resources.

The yield in production is significantly lower than the MNP. This phenomenon should be explained by the fact that the farm is not used to the maximum for the formation of the crop. The reason for this is the imperfect weather forecast, violation of agricultural techniques, the level of production organization, the presence of diseases, pests and weeds in crops and plantings of agricultural crops.

The main task of crop forecasting is to bring the potential yield closer to the actual possible yield and vice versa. For this, it is necessary to coordinate the improvement of plant requirements with environmental conditions. In addition, the economic efficiency of the measures used should be taken into account.

The quality of the harvest should not be assessed by its absolute value, but by the difference between the MMR and the PM. This difference is the amount of harvest that is lost due to the failure to use the potential opportunity for yield increase. The efficiency of harvest programs is higher, the closer the actual possible harvest is to the potential harvest. Under ideal conditions, the potential harvest should be equal to the actual possible harvest.

1.2. The influence of irrigation and fertilizers on the yield of corn crop.

Compliance with the optimal irrigation and feeding regime conditions significantly increases the yield and quality of corn products.

Intensive irrigation gives a particularly great effect when the plant is provided with the necessary amount of moisture in different phases of growth and development, which contributes to the complete absorption of nutrients.

The connection between the two factors, irrigation and irrigation, is carried out by means of the soil absorption complex, along with the exchange of cations, the soil is the main regulator of the concentration of solutions and the process of plant nutrition.

This process proceeds best when there are sufficient amounts of migrating potassium cations in the soil, the concentration of cobalt cations in the soil absorption complex has a positive effect on the soil structure, water permeability and other specific properties.

The main factors expressing soil fertility are the active balance of humus and the ion balance indicator in the absorption complex [8].

In the case of irrigation and fertilization, equal conditions are created for the plant throughout the entire vegetation period, which is especially important in the critical phases of its growth and development.

The movement of ions from the soil layers to the roots of the plant occurs mainly in two ways, by water flow or by independent movement of ions, which is caused by the action of electrokinetic forces. For example, sulfate and nitrate ions are characterized by weak bonds with soil mineral particles, and therefore their concentration in the soil solution is quite high. Through the soil solution, they come into contact with plant roots. The more intense their movement and displacement, the more complete the plant's needs are met.

Molybdenum and calcium ions are easily transported by water, while the displacement of potassium ions is insignificant. Phosphorus ions are poorly transported by water, which is why the degree of its absorption is determined by the interaction between plant roots and soil particles.

Special data of recent years clearly indicate the high efficiency of fertilizing irrigation. Such irrigation technology is successfully used for growing vegetable crops and flowering plants.

The introduction of mineral fertilizers into the soil together with water, especially during the critical phase of plant growth and development, leads to increased leaching, promotes the economical use of irrigation water and nutrients.

A number of scientists believe that the use of large amounts of fertilizers has an undesirable effect on soil properties: it disrupts its structure, changes the chemical composition of the soil and sharply reduces the effect of leaching.

Therefore, a great effect can be obtained as a result of leaching in the case when the irrigation process is adapted to the physical properties of the soil and fully provides the water demand of the plant in different phases of vegetation.

Improper irrigation leads to a violation of the water-air regime in the soil and creates conditions for the formation of anaerobiosis, which contributes to the formation of soluble and insoluble sulfates. The latter stops nitrification, the activity of bacteria and significantly reduces the ability of plants to absorb nutrients [9].

In the case of an optimal soil water and air regime, which is achieved through proper irrigation and loosening, relatively favorable conditions are created on soils with a neutral reaction for the activity of beneficial microflora, processes are accelerated that have a positive effect on the formation of active humus, nitrogen mineralization, and contribute to increasing soil fertility.

The introduction of mineral fertilizers enhances the growth and development of irrigated crops.

Mineral fertilizers affect the activity of microflora, as they supply it with nutrients (nitrogen, phosphorus) that are necessary for the life of soil microorganisms.

The concentration of the soil solution can be regulated by conducting the irrigation process in such a way as to ensure the plant's needs for both water and nutrients.

Optimal fertilization can significantly change the water balance and reduce water requirements. Growing field crops without fertilization or applying fertilizers in small quantities leads to a significant increase in water requirements [22].

In case of high soil fertility, irrigation is very effective, but the application of large doses of mineral fertilizers during optimal irrigation or the presence of high soil moisture against the background of optimal irrigation negatively affects the quantity and quality of the obtained products. Obviously, with high humidity and insufficient fertilization, the concentration of the soil solution decreases. This is associated with the leaching and loss of nitrate nitrogen, which is caused by an insufficient amount (frequency) of irrigation. An increase in the concentration of the soil solution inhibits plant growth and development, reduces crop yield. Therefore, in order to achieve high efficiency as a result of normal irrigation and irrigation of field crops, it is necessary to coordinate them in connection with their needs.

Based on the data from the experiments conducted on the corn variety ``Ajameti Tetrize" in the Saguramo field, it was concluded that one irrigation without fertilizing increases the yield by 20-25%, one fertilizing without irrigation increases the yield by 10-15%, and the simultaneous use of irrigation and fertilizing increases the yield by 80-100%.

In the case of an irrigation regime with optimal parameters, the corn yield increases progressively, while the water demand does not increase proportionally to the yield, as a result of which the water consumption per unit of production is lower.

Chapter II.
Agro-Amelioration Characteristics of the Mukhran-Saguramo Valley

2.1 Agro-climatic Factors of the Study Area

The majority of the field stationary experiments of the research results presented in the monograph were carried out at the Mukhrani Educational and Experimental Farm of the Georgian Agrarian University, which geographically belongs to the Mukhrani-Saguramo Plain. It represents the continuation of the Kartli Lowland and is located in its southeastern part, on the left bank of the Mtkvari River, in a northwestern direction, 45 km from Tbilisi.

According to the classification of I. Gachechiladze and K. Kelenjeridze, the climate of the valley belongs to the less continental type. The share of the continental factor in its formation is 60%, and the subtropical factor is 40%.

The Mukhran-Saguramo Valley has the shape of a triangle, the contour of which is limited on all sides by a chain of short mountain ranges and mountain ranges. The valley is slightly convex in the central part, characterized by a weakly expressed wavy surface and a general slight slope in the southeast direction. The valley is crossed by the valleys of the Ksani, Aragvi rivers and the latter's tributary, the Narekvav. To the south, the valley is bordered by the Skhaltbi and Saguramo watershed ranges, which separate the Mukhran Valley from the Mtkvari Valley. The bordering valleys to the north represent a branch of the Main Caucasus and act as watersheds for the Glekhura, Ksani, Narekvav and Iori rivers.

By processing long-term meteorological data, it has been established that the Mukhran-Saguramo Valley is characterized by a high thermal regime and a fairly long period of active vegetation.

In the eastern part, the Mukhran Valley is crossed by the Aragvi River, which flows from the north to the south with its tributaries, and in the west by the Ksani River.

The climate of the Mukhran Valley is greatly influenced by the Kartli Range, which is the main weather-dividing range. In terms of the continuous circulation of air masses, the Saguramo Ranges intercept moisture-bearing masses from both the west, from the Black Sea, and the east, from the Caspian Sea. Forced upward currents of air masses on the slopes of these ranges contribute to the intensification of condensation and, in general, rain-forming processes. Relatively higher amounts of precipitation fall on the slopes adjacent to ridges, especially on the slopes of mountainous ranges, than on the plain itself.

In general, the Mukhran-Saguramo Valley lowland zone is characterized by a straight and relatively calm relief, at an altitude of 500-700 m above sea level. Its location creates favorable conditions for the intensive development of irrigated agriculture. As for the territories with high elevations, they are characterized by the complexity of the relief and, obviously, the effective use of the area is limited without additional special hydrotechnical measures.

For a complete picture of meteorological conditions, we considered it necessary to use not only the Marto Mukhran, but also the multi-annual data of the neighboring meteorological station. The average annual amount of atmospheric precipitation on the entire Shida Kartli Plain (in its individual points) does not exceed 544 mm (Table 2.1.1).

Table 2.1.1

Amount of atmospheric precipitation according to average multi-year datamm)

Months / Districts	January	February,	March,	April	May	June	July	August	September	October	November	December	Total during the year	During the growing season
Kareli	33	34	36	49	75	61	43	34	42	47	43	39	536	300
Gori	31	32	34	46	69	56	40	32	39	43	40	35	498	282
Kaspi	17	20	23	44	87	63	44	35	39	31	28	19	450	312
Mtskheta	22	25	29	56	113	80	56	44	49	40	36	24	544	398
Mukhran	15	21	29	56	87	74	51	35	43	39	39	23	512	385
Medium	23	26	30	50	83	66	46	36	42	40	35	28	505	334

In this regard, the situation is relatively better in Mtskheta and other regions of Kartli. However, overall, there is no significant difference in the amount of precipitation between them. The largest amount of precipitation falls in May-June, and in the following months, including November, it varies within 30-55 mm depending on the region. The distribution of precipitation on the plain is uneven.

The Mukhrani Valley is not characterized by an abundance of precipitation, as a result of which droughts are quite frequent. In general, severe droughts in the region are observed in the summer and the first half of autumn and last on average 120-130 days. The reasons for long-term droughts should be considered high temperature regime, dry winds, and the fairly frequent spread of wind gusts, which cause drying of the soil.

The Shida Kartli plain is dominated by western winds. The main reason for this is the invasion of cold air masses. The wind speed does not exceed 3.0 m/s at all, although sometimes it reaches 15.0 m/s. The frequency of windy days varies within 70-90, and strong winds occur on 8-10 days during the corn growing season.

One of the main reasons for the occurrence of strong winds is narrow valleys, the direction of which coincides with the direction of the wind, and therefore they play the role of an aerodynamic funnel.

The Saguramo-Mukhrani Valley is also subject to the invasion of hot winds from the east, which causes intensive evaporation of water from the soil surface. The frequency of precipitation and winds in spring leads to the development of water and wind erosion processes of the soil, which leads to undesirable results in crops. As mentioned above, summers on the Shida Kartli Plain are hot and dry, and winters are cold.

We can judge the air temperature of the region by the data given in Tables 2.1.2 and 2.1.3.

The average temperature in January is –(0.5÷1.7) C, the average temperature of the summer months varies within +(18÷23) C. The maximum air temperature reaches +(39÷40) C, while the minimum temperature drops to –(20÷30) C.

Due to the relief and orographic features of the valley, cold air masses invade in spring and autumn, the late winter drought is reflected in the first decade of April, and the early autumn drought begins in the second half of October. According to long-term climatic data, severe droughts are quite rare in the territory of the zone. The number of frosty days varies within 100-120 days per year.

Table 2.1.2

Temperature regime indicators According to average multi-year data (^{0}C)

Months / Districts	January	February,	March,	April	May	June	July	August	September	October	November	December	Medium
Kareli	-1,7	0,3	4,3	9,5	14,8	18,2	21,3	21,5	17,3	11,6	5,5	0,3	10,2
Gori	-1,2	0,2	4,8	10,3	15,7	19,1	22,2	22,3	18,0	12,3	6,0	0,9	10,9
Kaspi	-0,5	0,6	5,4	10,7	15,8	19,7	23,1	23,2	15,9	13,0	6,4	0,7	11,4
Mtskheta	-1,1	0,5	4,9	10,2	15,5	18,9	22,1	22,0	17,7	13,2	6,1	1,0	10,8
Mukhrani	-1,7	0,1	5,6	9,8	15,2	18,6	21,8	21,7	17,5	12,0	5,9	1,0	10,0

Table 2.1.3

Average soil surface temperature according to multi-year data (^0C)

Months / Districts	January	February,	March,	April	May	June	July	August	September	October	November	December	Medium
Kareli	-2,2	-1,0	5,2	12,4	19,2	24,3	26,8	26,3	20,3	12,5	5,3	-1,2	12,3
Gori i	-2,0	-1,2	6,3	11,8	19,3	23,7	27,2	25,9	20,5	12,2	6,2	-0,3	12,6
Kaspi	-1,9	1,0	7,8	11,2	20,3	22,9	27,4	27,8	25,2	12,8	5,9	-1,1	13,1
Mtskheta	-1,4	1,2	2,2	12,8	20,5	24,5	28,0	27,2	21,5	13,8	6,2	-0,8	13,0
Mukhrani	-1,2	1,1	2,3	12,3	20,2	24,3	28,2	27,8	22,8	13,4	6,3	-0,7	13,4
Medium	-2,14	0,2	4,75	12,1	19,9	24,0	27,52	26,64	21,86	12,5	6,0	-0,82	13,0

As can be seen from Table 2.1.3, the lowest temperature is observed in January, minus 2°C, the highest – in July, plus 28°C. The period of positive temperature covers 180-220 days. It begins in mid-April and ends at the end of October.

The average value of active temperatures in the Saguramo-Mukhrani Valley is less than 4000°C and varies within 3245-3713°C.

In general, the temperature regime of the entire plain allows for a high corn harvest.

2.2 Composition of the humus cover of the test plot

The Mukhran-Saguramo Valley is an erosional plain, in the formation of which the main role was played by the material brought by the rivers Ksani, Aragvi, Narekvav and many other streams and flows in the form of pebbles and gravel. Clay soils are mainly distributed in the catchment areas of the Narekvav River, and pebble-stony soils in the remaining areas. The surface of the entire territory is gradually covered with carbonate-rich silty material, the thickness of which reaches 3-6 m. The soils of the Mukhran Valley have been studied by M. Sabashvili [19], V. Lataria, G. Talakhadze [22], I. Chkhenkeli [31] and others.

The following soil types are common in the study area: alluvial, brown and meadow-brown, mountain-meadow soils and their various subtypes. The majority of agricultural crops are grown on alluvial and meadow brown soils. According to the studies conducted by L. Guntaishvili [6], J. Machavariani [18], G. Talakhadze [22] and others, all types of alluvial-carbonate soils are common in Shida Kartli, among which silty soils occupy the largest area. According to the content of nutrients, they are fertile soils, but they are characterized by lack of structure and insufficient hydrophysical properties.

In this zone, brown meadow soils are also widespread, which are located mainly in the elevated, foothill and hilly areas of the plain.

The reaction of the upper layers of these soils is characterized by weak acidity, and the lower layers are characterized by neutrality. The amount of humus is (4.5-9.8)%. Its content decreases with depth and at a depth of 0.7-0.8 m it drops to 2%.

The nitrogen content of brown soils in the upper layers of the same depth is twice as high as in loamy soils. The distribution zone of brown soils is densely populated due to their favorable agro-industrial and climatic conditions. A large part of the territory is occupied by agricultural plantations. Of the natural plants, oak, hornbeam species and xerophytic shrubs are found here and there, as well as meadow plants on brown and meadow brown soils. The main factor in increasing yields is the provision of water, which should be provided by using optimal irrigation norms and regimes for agricultural crops. As we have mentioned, these soils are developed on sedimentary carbonate rocks of secondary origin of the secondary period. They are characterized by brown sheaf, a poorly differentiated genetic profile, and lime accumulation in the alluvial horizon.

A typical cross-section is provided below to morphologically characterize the soils of the field trials.

0–16 cm – brown, granular-grained structure. There are individual roots and stones, compacted structure, porous, clayey. It is fermented, covered with bark, and breaks up into layered sheets as a result of irrigation.

16–32 cm – light brown, banded, dense, heavy clayey, individual barks, shale.

32–48 cm – light straw-colored, granular-clayey, compacted, heavy clayey, with cracks, aggregates are shiny when cut. Individual roots are covered with starchy spots.

48–64 cm – light straw-colored, structureless, loose, loess-like clayey, stony, strongly shale.

According to the mechanical composition (Table 2.2.1), the soils of the test plot belong to heavy clay.

Table 2.2.1

Mechanical composition of soil

Depth, m	Fraction size in mm and their weight percentage content						
	1,0-0,25	0,25-0,05	0,05-0,01	0,01-0,005	0,005-0,001	0,001	>0,01
0,0-0,16	1,22	32,30	15,0	26,00	10,75	15,73	48,52
0,16-0,32	1,25	20,15	30,15	15,75	12,15	20,55	51,55
0,32-0,48	1,50	25,50	29,20	15,05	10,55	18,50	56,20
0,48-0,64	1,75	27,50	27,08	16,50	13,00	14,17	56,33

can be seen from the table, the soils are heavy clayey. In the first horizon, the content of physical clay particles in the upper layer is 48.52%, and increases towards the depth and reaches 56.33%. Soil compaction is caused by the shift of the clay fraction in depth, which in turn is due to the use of incorrect soil compaction and irrigation technology. Analysis of the water-physical properties of the soils of the test plot shows (Table 2.2.2) that they are characterized by an increase in bulk density with depth and, accordingly, a decrease in porosity. The marginal water holding capacity also decreases with depth, which should be explained by the presence of capillary hysteresis.

Table 2.2.2

Hydro-physical properties of brown meadow soils of the experimental plot of the Mukhrani training farm

Sampling depth, cm	Bulk density, g/cm^3	Specific gravity g/cm^3	Total porosity, %	Maximum molecular capacity %	Minimum Capacity, %	Filtration coefficient cm/sec
0-16	1,31	2,69	50,20	15,50	33,47	0,00245
16-32	1,43	2,72	47,50	14,10	30,01	0,00013
32-48	1,42	2,74	48,20	14,60	29,51	0,00001
48-64	1,45	2,75	47,30	14,50	29,21	0,00001
0-64	1,40	2,72	48,30	14,67	30,64	

Analysis of the hydro-physical properties of these soils shows (Table 2.2.2) that they are characterized by high bulk density, which increases significantly with depth and reaches 1.45 g/cm in the 0-16 cm layer from 1.31 g/cm in the lower 48-64 cm horizon. The total porosity, on the contrary, decreases with depth from 50.2 to 47.3%.

As for the minimum water holding capacity, its value decreases from the upper layers (33.47%) to the lower horizons to 29.21%.

The main data on soil agrochemical indicators show that at a depth of 0.6 m of the profile (Table 2.2.3) humus is evenly distributed, although it decreases slightly with depth, from 2.9% to 2.7%.

In accordance with the humus content in the soil, the content of soluble and hydrolyzable nitrogen is low, which indicates insufficient provision of nitrogen to plants. The content of total phosphorus is also low. The amount of soluble phosphorus varies from 0.11% to 0.44%, and easily soluble - from 2% to 2.07% per 100 grams of soil.

Table 2.2.3

Main agrochemical parameters of soil

Depth, cm	PH In water suspension	$CaCO_3$	humus % (by the Tyurin method)	nitrogen		phosphorus		Potassium	
				Total, % (Ginzburg method)	From hydrolysis, mg/100 g (by the Tyurin method)	Total, % (Ginzburg method)	From hydrolysis, mg/100 g (by the Tyurin method)	Total, % (Ginzburg method)	From hydrolysis, mg/100 g (by the Tyurin method)
0-20	8,1	9,1	2,9	0,16	11,6	0,11	2,80	1,20	54,5
20-40	8,2	9,3	2,8	0,17	12,11	0,30	2,30	1,20	50,87
40-60	8,2	9,3	2,7	0,18	12,29	0,44	2,07	1,10	50,70

The soils are not distinguished by a high exchangeability of potassium. It is evenly distributed in the profile and its total amount reaches 1.1-1.2%, only exchangeable potassium - 50.7-50.5 mg per 100 g of soil.

The content of carbonates in the soil varies on average within 9.1-9.3% and their slight increase with depth is observed. The soil has an alkaline reaction and changes insignificantly with depth.

2.3. Agro-ameliorative characteristics of corn

It is known that a number of prominent scientists, A.R. Konstantinov [55], I. Chkhenkeli, G. Tugushi [23], A.F. Ivanov [51] and others, at different times, put forward many criteria for assessing water requirements. These criteria, despite their different names such as `water supply coefficient', `water balance coefficient', `hydrothermal coefficient', `water deficit capacity coefficient', `evaporation coefficient', `dryness coefficient' or `dryness radiation index', `aortic deficit', `water-heat supply coefficient', etc., are mainly derived from the thermodynamic equilibrium conditions between the vapor pressure in the air and the soil moisture content.

It should be noted that the assessment and zoning of local natural climatic conditions is carried out according to various mutually determining factors (atmospheric precipitation, surface runoff, evaporation, radiation, air circulation, etc.).

The criteria are mainly adopted on the basis of generalization of specific data of climatic conditions of a particular region, and therefore the differences between them are quite understandable.

A number of scientists quite rightly point out that there cannot be a comprehensive, universal criterion that can fully take into account the role of all factors participating in the formation of the process and the resulting requirements of practice.

-In land reclamation practice, two characteristics are most often used simultaneously: the moisture coefficient and the evaporation deficit. For example, the studies of V.A. Surin, B.G. Shtepa, V.F. Nosenko and others have established that in irrigated agricultural areas characterized by a moisture coefficient of less than 0.33 and an evaporation deficit of more than 500 m3/ha, the use of artificial precipitation irrigation is not justified due to the large amount of water applied to the aquifer.

In the above-mentioned literary sources, the ratio of the annual total of atmospheric precipitation to the annual evaporation is considered to be an indicator of precipitation, while the evaporation deficit is the difference between evaporation and productively used atmospheric precipitation during the vegetation period.

The first is a qualitative indicator, since the lower the moisture content, or water supply coefficient K, the greater the water deficit compared to the demand under given specific climatic conditions.

The second characteristic ($E_0 - P$) – is a quantitative indicator of the deficit of aeration. The higher its value, the more water flow is required.

In our opinion, for the determination of the water supply of the region, it is expedient and correct to use the complex unified coefficient recommended by Professor G. Tughushi [23], which, compared to other analogous criteria, most fully reflects the true physical picture of the phenomenon.

The complex coefficient K_t is determined by the following formula:

$$K_t = \frac{E_0(P - E_0)}{P} \qquad (2.3.1)$$

where E_0 is the evaporation during the accounting period, which is calculated using the well-known formula of I. Ivanov [50]:

$$E_0 = 0{,}018(t + 25)^2(100 - r) \text{ მმ/თვე}, \qquad (2.3.2)$$

where P- is the amount of atmospheric precipitation during the reference period;

t – Average monthly air temperature;;

r– Average monthly relative air humidity.

The greater the absolute value of the complex K_t coefficient, the greater the tension caused by the lack or abundance of atmospheric precipitation. A positive sign of the coefficient indicates abundance, and a negative one indicates deficiency.

To illustrate the physical and industrial nature of this coefficient, it can be seen from the specific examples discussed by Prof. G.E. Tugushi [24] that in the case of the same deficit of atmospheric precipitation, a more tense situation is present where the water deficit coefficient K is smaller, and in the case of the same K, a more tense situation is present where the deficit of precipitation is greater in absolute value.

Similar conclusions can be drawn in the case when the area is characterized by abundant precipitation. One of the main goals of land reclamation measures is the normalization of water supply, that is, moistening (equalizing) the value of the water supply coefficient to one, and the complex coefficient K_t shows the fraction of the precipitation deficit that occurs per unit value of K. Thus, the K_t coefficient expresses the degree of difficulty in normalizing the existing natural conditions in connection with the regulation and optimization of water resources.

So, for example, if $P = E_0$,Then $\Delta P = P$, $K = 1$, While $K_t = 0$. This means that the normalization of natural conditions does not present any difficulties. To determine the complex coefficient as a generalized criterion, multi-year hydrometeorological data from three regions, Senaki, Mtskheta (Mukhran) and Telavi, which differ sharply from each other in terms of natural climatic conditions, were processed.

To assess the validity of the coefficients, the following indicators have been determined for the entire year, vegetation (April to September) and summer (June, July and August) periods, based on meteorological observation data:

ΣP -Total precipitation during the reporting period, mm;

ΣE_0 -Total evaporation during the same period, mm;

ΔP – ატმოსფერული ნალექების დეფიციტი, მმ;Atmospheric precipitation deficit, mm;

K – Water supply efficiency coefficient.

Water supply efficiency coefficient. ΣP , ΔP და K_t

The actual years of corresponding provision were determined separately. Based on the provision, we adopted the following classification: 5% – humid, 50% – average (median), 75% – average dry and 95% – dry years.

To determine the provision of indicators, we used the well-known relationship:

$$a = \frac{m - 0,3}{n + 0,4}, \qquad (2.3.3)$$

Where m- is the variable ordinal number of the member of the decreasing variational series;

n – the total number of members of the series.

The results of the calculations performed on the Mukhran-Sagu-Ramo example are given in Table 2.3.1.

As can be seen from the table, various indicators (ΣP , ΔP end K_t) The actual years of the same provision determined by the values coincide in some cases, and in some cases not. For example, the wet and average wet years turned out to be the same according to all indicators. The average (median) year turned out to be different according to each indicator. It is noteworthy that ΔP end K_t-s The median precipitation amounts are close to each other according to: $\Delta P = 517$ mm end $\Delta P = 513,6$ mm, which is also close to the multi-year arithmetic meanP_{Medium}m=526.2 mm.

Table 2.3.1

Various water supply – 5%, 25%, 50%, 75% და 95%, Determining real years ΣP, ΔP end K_t According to the magnitude of the indicators end ΣP, ΣE_0, ΔP, K_t Relevant climatic and melioration characteristics According to data from the Mukhrani meteorological station

Year category		Reporting period	Indicators corresponding to various collaterals ΣP By size				
Provision %			ΣP mm	ΣE_0 mm	ΔP mm	K mm	$K_{ტ}$ mm
1		2	3	4	5	6	7
ist	5%	Annual data	700,4	754,79	-54,38	0,92	-59,10
Moderately humid	25%		603,9	812,64	-208,74	0,74	-282,00
Median	50%		517,0	890,03	-373,00	0,58	-643,10
Average dry	75%		444,6	906,57	-461,97	0,49	-942,80
Dry	95%		383,3	912,97	-529,67	0,42	-1261,0
Moist	5%	Vegetation period data	482,9	580,22	-97,32	0,83	-117,25
Moderately humid	25%		426,4	555,32	-128,92	0,76	-169,63
Median	50%		327,4	665,63	-338,23	0,49	-690,26
Average dry	75%		240,7	574,36	-333,66	0,41	-813,80
Dry	95%		218,7	701,01	-482,31	0,31	-1555,80

Table 2.3.1 (continued)

1		2	3	4	5	6	7
Moist	5%	Summer period data	265,7	337,01	-71,31	0,78	-91,42
Moderately humid	25%		179,4	319,81	-140,47	0,56	-250,83
Median	50%		149,6	386,24	-236,64	0,39	-606,76
Average dry	75%		130,9	416,83	-285,93	0,31	-922,35
Dry	95%		81,1	425,67	-344,57	0,19	-1813,50
			Indicators corresponding to various collaterals **ΔP** By size				
Moist	5%	Annual data	The same as **ΣP** According to				
Moderately humid	25%		The same as **ΣP** According to				
Median	50%		397,5	776,13	-338,63	0,51	-663,98
Average dry	75%		The same as **ΣP** According to				
Dry	95%		332,9	993,56	-600,66	0,35	-1716,10
Moist	5%	Vegetation period data	469,3	565,47	-96,17	4,87	-19,74
Moderately humid	25%		399,7	598,52	-198,82	0,66	-301,24
Median	50%		The same as **ΣP** According to				
Average dry	75%		180,2	581,20	-397,00	0,31	-128,00
dry	95%		231,0	792,04	-561,04	0,29	-1934,00

Table 2.3.1 (continued)

1	2	3	4	5	6	7
Moist 5%	Summer period data	270,6	344,35	-74,25	0,79	-94,00
Moderately humid 25%		168,5	355,55	-187,06	0,47	-398,00
Median 50%		The same as ΣP According to				
Average dry 75%		The same as ΣP				
Dry 95%		154,2	415,64	-361,44	0,002	-1807,20
		Indicators corresponding to various collaterals according to the magnitude of K				
Moist 5%	Annual data	The same as ΣP და ΔP According to				
Moderately humid 25%		The same as ΣP და ΔP According to				
Median 50%		513,6	873,55	-359,95	0,58	-620,00
Average dry 75%		409,4	857,20	-447,80	0,47	-952,00
Dry 95%		The same as ΔP According to				
Moist 5%	Vegetation period data	The same as ΔP According to				
Moderately humid 25%		470,1	663,78	-193,68	0,70	-276,50
Median 50%		307,4	650,10	-342,90	0,47	-729,00
Average dry 75%		The same as ΔP According to				
Dry 95%		The same as ΔP According to				
Moist 5%	Summer period data	The same as ΔP According to				
Moderately humid Average dry 25%		The same as ΔP According to				
Median 50%		The same as ΣP და ΔP According to				
Average dry 75%		The same as ΣP და ΔP According to				
dry 95%		The same as ΔP - According to				

As for ΔP According to the indicator, the amount of precipitation is $\Sigma P = 397{,}5$ mm, Which is clearly reduced compared to the above and accordingly indicates that ΔP In this case, it does not provide an objective assessment of the current situation.

Average dry year ΣP end ΔP According to the indicators, they are the same, but according to K_t, they are different. In this case, it is interesting that almost all the indicators of the yea r (ΣP, ΣE_0, ΔP, K) It is slightly less than the same indicators of the previous year and is proportional to each other. This indicates that the final results, in all cases, do not differ significantly from each other, but the results obtained according to K_t are characterized by a relatively high level of reliability.

ΔP end K_t The dry years determined by the ΔP- From a dry year determined according to.

Conducting a similar analysis for the vegetation and summer periods gives the same picture. In all cases, the determination of the categories of years by the K_t indicator according to the water supply for each month is more reliable than ΣP end ΔP-Years determined by.

It should be noted that the results determined according to K_t coincide four times.

ΣP-s, Eleven times – ΔP-s and only four times different from the results obtained according to the above indicators. AndΣP end ΔP-s The results determined by the method coincide only 6 times. This indicates that for determining true water scarcity

ΣP-Represents a more objective ΔP-compared to, and K_{tt} – even more objective than both of the above indicators. Such a conclusion was also confirmed in the analysis of similar plots of multi-year meteorological data for the Senaki and Telavi districts. At the same time, it was also revealed that Kt Machvenebeli is the best criterion for the climatic and melioration regions of the territory. This is confirmed by comparing the average and median multi-year indicators of the Mtskheta (Mukhrani), Telavi and Senaki municipalities located in different climatic zones (Table 2.3.2).

The assessment of the water supply of the above-mentioned regions according to the Kt criterion shows the following: the deficit of atmospheric precipitation in the Mtskheta and Telavi municipalities does not differ significantly from each other (359.95 mm and 385.85 mm), but the difference according to the K_t criterion is significant (-620.6 mm and -431.25 mm). This means that in terms of annual amount, the Mtskheta municipality is much less provided with atmospheric precipitation than the Telavi municipality, which is a generally known fact.

The K_t criterion of the Senaki municipality is positive, which indicates that the mentioned municipality is characterized by excess natural precipitation.

A comparison of the vegetation periods shows that in terms of absolute magnitude, the deficit of atmospheric precipitation in Mtskheta Municipality is much greater than in Telavi Municipality (349.20 > 269.50).

Table 2.3.2

Comparison of water supply indicators of Mtskheta (Mukhrani), Telavi and Senaki districts located in different climatic zones, determined according to average-median multi-year data and their corresponding other climatic ameliorative characteristics – ΣP , ΣE_0 , ΔP , K_t

Districts	Period	ndicators corresponding to 50% coverage ΣP According to the size of				ndicators corresponding to 50% coverage ΔP - According to the size of				ndicators corresponding to 50% coverage K_t- According to the size of			
		ΣP mm	ΣE_0 mm	ΔP mm	K_t mm	ΣP mm	ΣE_0 mm	ΔP mm	K_t mm	ΣP mm	ΣE_0 mm	ΔP mm	K_t mm
Mtskheta	Annual data	517,0	890,03	-373,00	-643,10	397,5	776,13	-338,63	-663,98	513,6	873,55	-359.95	-620.60
Telavi		695,6	984,04	-287,44	-404,84	685,8	1019,68	-333,88	-498,32	652,3	959,15	-306.85	-431.25
Senaki		1227,4	828,47	+898,93	+431,17	1656,9	895,02	+761,88	+411,82	1660,5	972,98	-687.52	+404.42
Mtskheta	Vegetation period data	327,4	665,63	-338,23	-690,26	The same as - ΣP-ს მიხედვით				307,4	650,10	-342,90	-729,57
Telav		488,4	634,28	-145,88	-189,45	501,30	710,20	-269,50	-299,28	The same - ΔP -as according to			
Senaki		799,0	510,99	+288,91	+185,20	732,40	466,55	+265,85	+169,33	The same - ΔP -as according to			
Mtskheta	Summer period data	149,6	386,24	-236,64	-606,76	The same as - ΣP -as according to				The same - ΔP as according to			
Telav		234,7	389,37	-154,67	-257,78	155,30	441,61	-286,31	-818,03	The same - ΔP -as according to			
Senaki		427,5	255,18	+172,32	+103,18	434,70	272,62	+168,08	+101,93	The same ΔP as according to			

Analysis of data for the summer period primarily shows that in Telavi Municipality ΔP The criterion is characterized more convincingly than ΣP criterion, since the reporting years determined by the aforementioned criterion coincide with each other.

Data analysis shows that summer in Telavi district is much hotter than in Mtskheta district, therefore the precipitation is also higher (441.61 > 386.24). This phenomenon well reflects the K_t indicator, Which in absolute terms is much higher in Telavi Municipality than in Mtskheta Municipality (818.03 > 606.76).

Thus, the analysis of actual data shows that the K_t indicator, in comparison with other indicators,

he difference between them (349.2 – 269.5 = 79.7 mm) can be considered as the difference between irrigation norms, which is only one irrigation norm – approximately 800 m^3/ha. But the indicator for Mtskheta municipalities is 2.4 times higher than for Telavisa (729.57 : 299.38 = 2.4), which means that the artificial regulation of water resources in Mtskheta municipality is 2.4 times more severe than in Telavi municipality.

(ΣP, ΔP, K) It clearly and truly expresses the quality of water supply, both for different annual supply periods and for climatic and geographical zones.

2.4. Climatic factors affecting corn yield

It is known that the yield of agricultural crops is closely related to the efficiency of using the potential capabilities of climatic factors, which in turn requires reliable determination of appropriate complex indicators.

Among the climatic factors, the following are particularly important: solar radiation, radiation balance, solar photosynthetic active radiation (PAR), air temperature, sum of air temperatures, atmospheric precipitation, etc.

It is also known that the calculation of solar radiation and radiation balance is based on meteorological observation data, which are obtained as a result of direct measurement of direct and diffuse radiation.

Direct and diffuse radiation belong to the short-wave part of the spectrum. The total total solar radiation income consists of direct and diffuse radiation.

At any given moment in time, there is a certain ratio between the incoming and outgoing radiant energy on the Earth's surface.

Based on our analysis, the difference between these energies, or radiation balance, can be expressed as follows:

$$R = Q_p + Q_g + E_{at} - Q_{an} - E_d, \qquad (2.4.1)$$

Where- Q_p end Q_g–Accordingly, there are direct and diffuse radiations;

Q_{an} -Reflected radiation;

E_{at} end E_d –– Accordingly, the long-wave radiation of the atmosphere and the Earth.

Depending on the ratio of income and expenditure, the radiation balance can be both positive and negative.

Based on the data of long-term observations of the Mukhrani meteorological station, we have established a relationship between the total solar radiation and the increasing sum of active temperatures.

$$\sum R = a\sum t > 10°C + b .\wedge \qquad (2.4.2)$$

Similar relationships also reflected the relationship between the increasing sum of the radiation balance and the increasing sums of radiation:

$$\wedge\sum R = \sum Q = d . \qquad (2.4.3)$$

(2.4.2) end (2.4.3) By combining the equations, we get:

$$\sum Q = \frac{a\sum t > 10°C + b - d}{c}, \qquad (2.4.4)$$

Where a, b, c, and d are the empirical coefficients of the equations

From literary sources [26, 27, 28] it is known that the shield accounts for 43% of direct radiation and 57% of diffuse radiation, i.e. approximately 50% of total radiation. Taking this into account, we can write:

$$\text{far} = \frac{a\sum t > 10°C + b - d}{c}. \qquad (2.4.5)$$

We can also represent the shield as a function of the radiation balance:

$$\text{far} = \frac{1}{2}\sum Q = \frac{\sum R - d}{2c}. \qquad (2.4.6)$$

These issues, as well as water supply issues, have been worked out for years at the Department of Hydraulic Engineering and Land Reclamation of the Agrarian University under the leadership of Prof. G.E. Tughushi.

The results obtained for different meteorological years are given in Table 2.4.1 and illustrated in the appropriate graphs (Fig. 1 and Fig. 2). The estimation of the actual and analytical values does not exceed 11%, and the standard deviation is 8%.

Table 2.4.1

Multiannual averaged data for estimating solar radiation and its dependent factors during the growing season for corn yield

#	Months, Indicators	IV	V	VI	VII	VIII	IX	X
1	2	3	4	5	6	7	8	9
1	Sum of active temperatures $t > 10°C$	374,4	508,8	588,9	689,0	659,3	530,1	368,6
2	*The total radiation balance is r*	6856,8	11962,5	12904,7	13178,0	12224,5	9216,4	5472,7
3	Total of the PAR	6040,6	7156,2	28119,8	8463,9	7213,1	5399,6	3893,1
4	The increasing sum of active temperatures $\sum t > 10°C$	374,4	883,2	1472,1	2161,1	2820,4	3350,5	3713,0
5	The increasing sum of the radiation balance $\sum R$	6856,8	18819,3	31724,0	44902,0	57123,5	66342,9	71815,6
6	The increasing sum of the PAR	6040,6	13196,8	41316,6	49780,5	56993,6	62393,2	66286,3

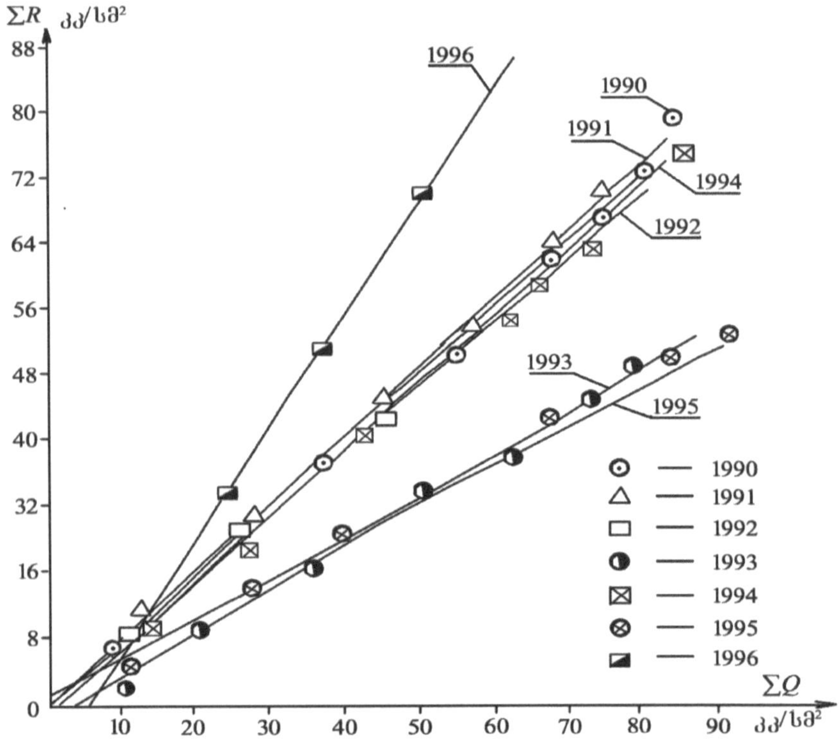

Fig. 1. Relationship between radiation balance and total radiation during the growing season

According to the conducted research, the solar radiation indicators during the vegetation period vary within the following limits. The sum of active temperatures (3245 3731)^{0}C, on average – 3713^{0}C. Radiation balance (46472 102651) cal/cm^2,irradiance – (4020- 46091) cal/cm^2, on average – 66286 cal/cm^2.

These indicators fully ensure the maximum potential yield not only of corn, but also of other agricultural crops widespread in the region. It should be noted that in some years the month of April turned out to be insufficiently warm, but instead the temperature

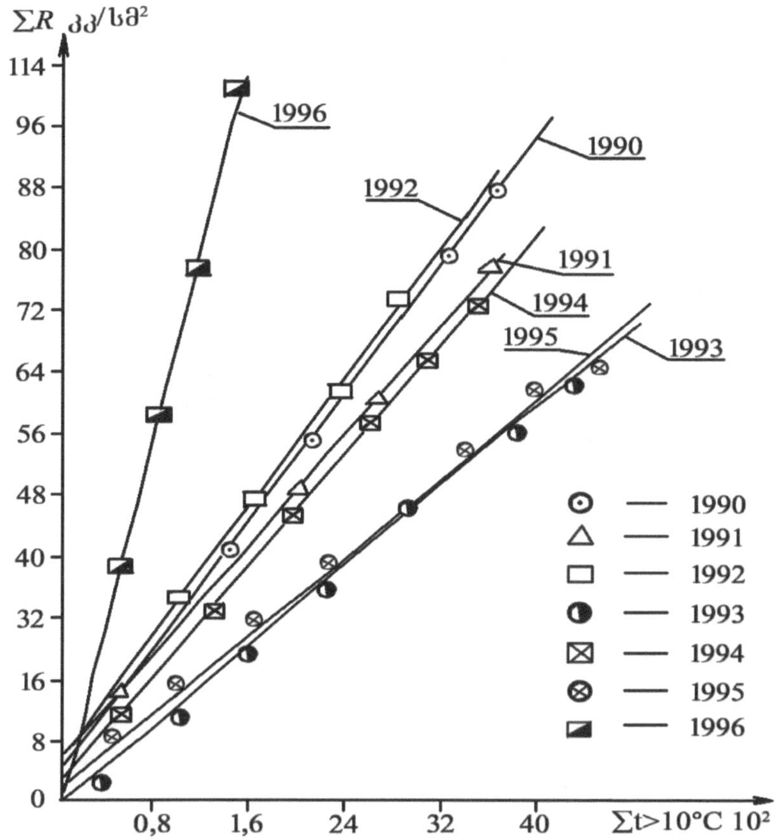

Fig. 2. Relationship between radiation balance and active temperature during the growing season

regime of the month of October ensured the constancy of the energy balance during the vegetation period. Despite this, the change of the phase of the thermal period led to a delay in the development of corn seedlings and, accordingly, a decrease in yield. In particular, such a phenomenon was noted in the third year of our field trials.

2.5. Meteorological indicators of the Mukhran-Saguramo field

In order to quantitatively assess the meteorological factors affecting the growth and development of plants, and in particular corn, the duration of individual phases, the irrigation regime and timing, observations were made throughout the vegetation period on atmospheric precipitation, absolute and relative air humidity, the dynamics of moisture in the active soil layer, etc. Based on a multifaceted analysis and appropriate processing of the observation materials, dependences were obtained for determining corn water consumption, taking into account meteorological and other factors, which are discussed in detail in the following chapters of this work.

The average values of the indicators given in Table 2.5.1 give an idea of the meteorological conditions in the years of conducting the experiments. The data in the table show that the air temperature and atmospheric precipitation differ more or less from the average multi-year data. It is noteworthy that in the first and fourth years the average air temperature in April and May was much higher than the average multi-year. The created temperature regime ensured normal germination, growth and development of corn, which was expressed in an increase in yield compared to other years. The third year, on the contrary, was distinguished by low temperatures in the months of April-May, which significantly affected the yield and to some extent caused its reduction.

Table 2.5.1

Atmospheric precipitation, air temperature and other indicators characteristic of meteorological conditions in the years of the experiments

Months / Meteorological data	IV	V	VI	VII	VIII	IX	Total atmospheric precipitation ΔP	Medium temperature $\Sigma t°C$	Evaporation ΣE_0	Atmospheric precipitation deficit $\Delta P - E_0$	*Water supply coefficientK_t*
1	2	3	4	5	6	7	8	9	10	11	12
Medium perennial											
Atmospheric precipitation	56	113	80	56	44	49	398,00				
Air temperature	10,20	15,50	18,90	22,10	22,00	17,70		17,73			
Evaporation											
First year											
Atmospheric precipitation	78,90	60,50	105,60	71,90	10,40	10,40	337,70				
Air temperature	11,36	16,50	18,70	21,40	23,36	19,20		18,42			
Evaporation									656,75	-219,00	-361,10

Table 2.5.1 (continued)

1	2	3	4	5	6	7	8	9	10	11	12	
	Second year											
Atmospheric precipitation	63,30	84,90	86,20	30,70	35,20	42,10	342,00					
Air temperature	10,90	15,86	19,70	23,80	21,26	16,80		18,05				
Evaporation	57,90	63,10	86,30	120,00	104,00	75,40			506,90	-164,80	-243,70	
	Third year											
Atmospheric precipitation	30,10	74,40	97,30	46,40	96,20	91,50	435,90					
Air temperature	12,80	16,50	18,63	22,96	20,80	16,70		18,06				
Evaporation									577,00	-141,10	-186,70	
	Fourth year											
Atmospheric precipitation	142,4	36,30	42,40	62,40	52,60	21,00	357,10					
Air temperature	10,93	14,66	18,93	22,73	20,03	19,60		17,81				
Evaporation									517,60	-160,51	-232,60	
	Fifth year											
Atmospheric precipitation	46,20	88,90	71,10	3,50	22,30	45,60	277,60					
Air temperature	12,63	17,18	19,67	22,30	25,50	17,56		19,09				
Evaporation	37,10	83,50	86,10	136,90	117,70	91,20			522,00	-274,40	-548,30	

In the remaining years, the characteristics of the temperature regime differed little from the average for many years, and therefore the yield fluctuated within the limits close to the average indicators.

Atmospheric precipitation was only once 38 mm more than the average for many years. In the same year, evaporation was also the greatest - 577 mm, while the deficit of atmospheric precipitation was 140 mm and the complex coefficient of water supply - 186.7, which was the smallest compared to other years.

The first year of the experiment was warm and humid, characterized by the smallest deficit of precipitation and high water supply, which determined the maximum yield. The data in the table also clearly shows that the fourth year was the most difficult, since the absolute value of the complex coefficient of water supply was the largest - 548 mm, and accordingly, the deficit of atmospheric precipitation was equal to 274 mm.

For a more complete assessment of climatic conditions in the years of conducting the experiments, multi-year meteorological data were processed, on the basis of which the following five categories of years were established in accordance with the provision of heat (average temperature) and precipitation during the vegetation period.

According to atmospheric precipitation:

1. Dry year, with 95% certainty, $\Sigma P = 180{,}96$ mm;
2. Average dry year, 75% coverage – $\Sigma P = 226{,}10$ mm;
3. Average (median) year, 50% coverage– $\Sigma P = 226{,}10$;
4. Average wet year, with 25% coverage – $\Sigma P = 226{,}10$;
5. Wet year, 5% collateral $-\Sigma P = 549{,}70$ mm;

By average temperature::

1. Hot year, with 5% collateral – $t_{saS} = 19{,}55°C$;
2. Average hot year, 25% coverage $-t_{saS} = 18{,}50°C$;
3. Average (median) year, 50% coverage $-t_{saS} = 17{,}76°C$;
4. Average cold year, 75% coverage $-t_{saS} = 17{,}53°C$;
5. Cold year, 95% guaranteed $t_{saS} = 17{,}01°C$.

The years of conducting the experiments are characterized by the following provisions:

- In the first year, the daily precipitation during the vegetation period is $\Sigma P = 337{,}70$ mm, With 50% of each. The average temperature in the same period, $t_{sas} = 18{,}42°C$ with 30% of each.

Thus, the year under consideration belongs to the average (median) year in terms of atmospheric precipitation, and in terms of temperature it is close to the average hot year;

- In the second year $\Sigma P = 337{,}70$ mm, 48% With security . $t_{sas} = 18{,}05°C$, 41% With security According to atmospheric precipitation, it approaches the average (median) year, and according to temperature, it is also close to the same year. This year can be considered as an average of the years in which the research was conducted, which was also reflected in the yield indicators. This will be explained in the following chapters;
- In the third year $\Sigma P = 337{,}70$ mm, 18% With security $t_{sas} = 18{,}06°C$, 40,50% With security. The atmosphere is approaching an average humid year in terms of precipitation, and an average (median) year in terms of temperature;
- in the fourth year $\Sigma P = 337{,}70$ mm, 38% With security, $t_{sas} = 17.81°C$, 48% With security. It approaches the average (median) year in terms of both precipitation and temperature;
- In the fifth year $\Sigma P = 237{,}60$ mm, 71% With security. $t_{sas} = 19{,}09°C$, 7% With security .

In terms of precipitation, it is approaching an average dry year, and in terms of temperature, it is approaching a hot year.

According to meteorological conditions, the above classification of years corresponds to the usual traditional concepts, but it should be noted that for a clearer and more unambiguous classification, we consider it appropriate to determine the categories of years according to their provision by the complex Kt coefficient of water provision proposed by us, which simultaneously takes into account both atmospheric precipitation and temperature and other factors of the external environment that are reflected in the aridity.

Then, according to the provision, we would have the following categories: dry (hot) - 95%, moderately dry (hot) - 75%, average (median) - 50%, moderately humid (cold) - 25%, humid (cold) - with 5% provision.

Chapter III

Predictable corn yield and water requirements

3.1. Corn water requirements and soil moisture dynamics

As the conducted studies have shown, one of the main factors in obtaining a corn crop is the provision of water to the plant, therefore, naturally, this issue requires proper study and multifaceted analysis.

The water demand of agricultural crops is determined by empirical, semi-empirical and theoretical approaches proposed by various authors. The method of A.M. Alpatyev [38] is still very popular, according to which the evaporation rate, or water absorption of the plant, is determined by the following simple formula:

$$E = KD7\ \text{১б}\ E = K\sum D, \qquad (3.1.1)$$

Where: E -is the water demand, mm;;

D – Air humidity deficit, mm;

$\sum D$ – Total deficit during the growing season, mm;

K – Water demand coefficient or bioclimatic coefficient, which is determined according to field experiment data, based on a balance calculation.

The amount of water actually evaporated (taking into account atmospheric precipitation) will be calculated based on the corresponding soil moisture at the beginning and end of a certain period of time using the following formula:

$$E = W_s + \mu P + m + W_b, \qquad (3.1.2)$$

Where: W_s end W_b – Accordingly, the estimated water supply (in the active soil layer) at the beginning and end of the period under consideration;

P – Atmospheric precipitation during the same period;

m – Amount of water supplied;

Active soil layer; μ – The precipitation interception coefficient, which is determined by the formula

$$\mu = \frac{100\gamma H\left(r_{zR} - r\right)}{P}, \qquad (3.1.3)$$

Where:

γ -is the volumetric mass of the soil;

H-Active soil layer;

r_{zR} -The marginal or minimum water holding capacity of the soil;

r – Soil moisture at the beginning of the growing season.

$\sum D$ -is determined by meteorological data, and

K – according to field experiments. Obviously, by combining these two parameters, E is easily determined.

In physical terms, K (water demand coefficient) represents the amount of water consumed per unit of air deficit. Experimental studies have established that the water demand of agricultural crops changes during the growing season.

According to the multi-year field experimental data conducted in the natural climatic conditions of Georgia, and in particular, in the Mukhrani Valley by the employees of the Department of Agricultural Hydrotechnics and Land Reclamation of the Agrarian University, the value of the coefficient in spring is 0.6, and in summer - 0.5. The values of the coefficient specified by Prof. I. Chkhenkeli [31] are given in Table 3.1.1.

Table 3.1.1

Values of the coefficient of water demand in relation to the air humidity deficit (D) according to experimental studies and calculation formulas of K

D, მმ	K	$K_{საანგ.}$	$\frac{K}{K_{saang.}} = \frac{E}{E_{saang.}}$
0,1	0,60	–	–
1,0	0,60	–	–
6,7	0,60	0,600	1,000
7,0	0,58	0,574	1,010
7,5	0,55	0,542	1,014
8,0	0,52	0,515	1,009
8,5	0,49	0,490	1,000
9,0	0,47	0,468	1,004
9,5	0,45	0,448	1,004
10,0	0,43	0,430	1,000
11,0	0,40	0,398	1,005
12,0	0,37	0,371	0,997
13,0	0,35	0,348	1,005
That is	That is	That is	That is
14,0	0,33	0,328	1,005
15,0	0,32	0,310	1,032

The mathematical processing of the data presented in the table reveals that there is a close functional relationship between the K coefficient and the air humidity deficit when K changes from 6.7 to 15, which is expressed as follows:

$$K = f(D) = 2.76D^{-0.807}. \qquad (3.1.4)$$

Let us represent this formula as follows:

$$K = 2{,}76D^{-0{,}807}\left(\frac{6{,}7}{6{,}7}\right)^{-0{,}807} = \frac{D^{-0{,}807}}{6{,}7^{-0{,}807}} \cdot 2{,}76 \cdot 6{,}7^{-0{,}807}$$,That is

$$K = 0.6\left(\frac{D}{6.7}\right)^{-0.807}. \qquad (3.1.5)$$

The approximate values of the K coefficient determined by this formula are given in the same table, from which it is clearly seen that the difference between the experimentally and analytically determined K values is practically insignificant.

From the above formula, the total water content can be determined by the following assumptions:

$$E = K \cdot D = 2{,}76D^{-0{,}807} \cdot D = \frac{0{,}6}{6{,}7^{-0{,}807}} D^{0{,}193}, \text{ That is}$$

$$E = 2{,}76D^{0{,}193}. \qquad (3.1.6)$$

From the above formulas it is clearly seen that with the increase in air humidity deficit, the water demand coefficient decreases, while the total water demand increases.

The water demand coefficient K also depends on the depth of the active layer, which is confirmed by the data of field-experimental studies conducted under the leadership of Prof. I.A. Chkhenkeli (Report of the Department of Agricultural Reclamation of the Agrarian University `Clarification of the method for correcting irrigation timing according to moisture deficit', I.A. Chkhenkeli, P.S. Sichinava, J.I. Gubeladze), the final results of which are given in Table 3.2.1.

As a result of mathematical processing of these data, it was found that the dependence of the K coefficient on the active H layer can be represented as the following function:

$$K = AH^m, \qquad (3.1.7)$$

Where the quality indicator $m = 1{,}04$.

Active layer H_1 end H_2 Matching of values K_1 end K_2 The values will be:

$$K_1 = AH_1^m$$
$$K_2 = AH_2^m. \qquad (3.1.8)$$

From these equations it follows that:

$$K_2 = K_1\left(\frac{H_2}{H_1}\right)^m.\wedge \qquad (3.1.9)$$

3.2. Corn yield forecasting and crop water demand

By modifying the above-mentioned A.G. Alpatiev [38] method and the corresponding relationship, the value of the water demand coefficient has been determined in accordance with the experimental data, which, from the point of view of practical application, is characterized by quite high accuracy.

Therefore, if we know the value of the coefficient K at any value of the argument H, then we can understand the value of the coefficient K when

Therefore, if we know the value of the coefficient K at some value of the argument H_1, then we can find the value of the coefficient K_2 when $H = H_1$; (3.1.9) According to the formula, if we take as a given $K = K_1$ The meanings that $H = H_1 = 0,6$ m, Then the search values when $H = H_2$, Definition by the formula::

$$K_2 = K_1\left(\frac{H_2}{0,6}\right)^{1,04}. \qquad (3.2.1)$$

Table 3.2.1

Values of water demand coefficient K according to the depth of the active soil layer and air deficit

Experimental data				According to the calculation formula			
D mm / *H* mm	6,7	10	15	*D* mm / *H* mm	6,7	10	15
0,6	0,60	0,43	0,32	0,6	0,60	0,43	0,32
0,7	0,60	0,50	0,37	0,7	0,60	0,504	0,363^
0,8	0,60	0,57	0,42	0,8	0,60	0,579	0,417

The values of the K coefficient calculated by this formula are given in Table 3.2.1. As can be seen from the table, the actual and formula-determined values of the K coefficient are in good agreement. If we insert the value of K_1 into the latter formula when H=0.6 m, we obtain:

$$K_2 = \left[0,6\left(\frac{D}{6,7}\right)^{-0,807}\right]\left(\frac{H_2}{0,6}\right)^{1,04}, \qquad (3.2.2)$$

Water demand is expressed by the formula:

$$E = K_2 D\left[0,6\left(\frac{D}{6.7}\right)^{-0,807}\right]\left(\frac{H_2}{0,6}\right)^{1,04} D = 2,76D^{0.193}\left(\frac{H_2}{0,6}\right)^{1,04}. \qquad (3.2.3)$$

One of the goals of our research was also the experimental verification and further generalization of this formula.

In order to determine the water requirements of corn, in the variants and their repetitions, in the conditions of different irrigation regimes (variants 1, 5, 8) developed according to the agronomic regulations, soil moisture was measured every 5-10 days with a hygrometer or by the method of drying samples. In addition, we received meteorological data from a meteorological station located near the experimental plot - (atmospheric precipitation, air and soil temperature), relative air humidity, air moisture deficit, etc. According to the meteorological data, the amount of water evaporated in a particular time interval (total evaporation) and the estimated irrigation dates were determined. Then we determined the active soil moisture, the actual date of irrigation, and the actual evaporation.

The dynamics of soil moisture under different irrigation regimes and in different years was determined in the form of balance tables, which showed the dates of the calculation interval, its duration in days and nights, the moisture content in the soil at the beginning and end of the irrigation period, in millimeters, in percent by weight, and in percent by maximum water capacity. In the first decade of the calculation period, the air temperature was low.

3.3. Corn Irrigation Dates and Fertilizer Doses Affecting Yield

The irrigation dates were adjusted by decade, taking into account the air temperature and precipitation. The actual water demand of corn is determined based on the balance calculation under different irrigation regimes.

Table 3.3.1 shows the dates and number of irrigations actually carried out in different years. As can be seen from the table, the number of irrigations varies depending on the meteorological conditions. Irrigation immediately after sowing, as a mandatory agrotechnical measure required for uniform germination, was not carried out in some years due to rains that fell immediately after sowing.

Table 3.3.1

Watering dates under different irrigation regimes

Years	Watering options	Post-sowing irrigation	Vegetation fleas	
			Before tassel emergence	After tassel emergence
1	0,7 0,6	– –	– –	22/VII; 8/VIII; 5/IX 18/VIII; 6/IX
2	0,7 0,6	24/IV 24/IV	11/VII –	11/VIII; 1/IIX 25/VII; 31/VIII
3	0,7 0,6	16/IV 16/IV	I5/VII –	18/VII 18/VII
4	0,7 0,6	– –	8/VI; 11/VII 1/VII	1/IX; 26/IX 23/VIII
5	0,7 0,6	26/IV 26/IV	8/VII 13/VII	26/VII; 13/VIII; 9/IX 11/VIII; 21/IX
Note: Irrigation was supposed to be carried out on August 22 of the third year, but it was not carried out because 65 mm of atmospheric precipitation fell.				

As can be seen from Table 3.3.1, the number of waterings varies in different years depending on meteorological conditions within the following limits:

- 2-4 in pre-watering humidity conditions of 70-65%, from here until the emergence of the caterpillars 0-2, and then 1-3;
- 1-3 in pre-watering humidity conditions of 60-55%, from here until the emergence of the caterpillars 0-1, and then 1-2.

Such diversity was caused by different meteorological conditions.

The second half of April of the first year was rainy ($P = 43{,}4$ mm) , In the second decade and $P = 26{,}5$ mm

In the third decade). At the same time, the air temperature was low in both the second and third decades. In the second decade $t_{saS} = 9{,}7°C$,

In the third decade $t_{saS} = 9{,}5°C$.Soil temperature in the sec-ond decade $t_{saS} = 12{,}3°C$, And in the third decade $t_{saS} = 11{,}3°C$.

The relatively high soil temperature in the second decade was due to the high temperature in the first decade. $t_{saS} = 13{,}8°C$. Corn was sown at the end of April, and irrigation was no longer necessary immediately after sowing. After sowing, the air temperature gradually increased.

In the second year, both air and soil temperatures were low, so corn was sown in the third decade of April (air temper-ature $t_{saS} = 13{,}3°C$,Soil – $t_{saS} = 12{,}9°C$).

After sowing, both air and soil temperatures progressively increased until the end of May. There was no drop in temperature.

In the third year, corn was sown on April 25. The third decade was rainy (P=38.5 mm). The first decade of May was also rainy.

In the third year, both air and soil temperatures were high in April. The air temperature in the first decade was $t_{sas} = 13{,}3°C$, In the second – $t_{sas} = 12{,}8°C$, In the third – $t_{sas} = 12{,}3°C$ .And the soil temperature in the first decade– $t_{sas} = 12{,}2°C$, In the second –$t_{sas} = 14{,}2°C$,In the third– $t_{sas} = 13{,}4°C$.

The amount of precipitation in the first decade was not large.– $P = 13$ mm, n the second – $P = 17{,}1$ mm, In the third – $P = 11{,}7$. All these circumstances led to early sowing of corn – at the beginning of the second decade, and accordingly, the harvest is high.

n the fourth year, both soil and air temperatures were high. The soil temperature in the second decade was $t_{sas} = 12{,}7°C$, air – $t_{sas} = 12{,}1°C$, but then, in the next decade, the air and soil temperatures dropped sharply $t_{sas} = 11{,}5°C$, and in the third decade, it was , which was accompanied by a large amount of atmospheric precipitation in the third decade of April – $P = 74{,}8$ mm, in the first decade of May – $P = 26{,}6$ mm. As a result, irrigation was not carried out immediately after sowing. The above-mentioned circumstance hindered the emergence of corn and part of the area was re-sown.

In the fifth year, the conditions were almost similar, with the difference that the average temperature of the first and second decades of April was lower compared to the third year – $t_{sas} = 11{,}1°C$ in the first decade, and in the second – $t_{sas} = 12{,}5°C$. The average temperature of April was also lower. Corn was sown at the beginning of the second decade. As is known, post-sowing irrigation is one of the important agrotechnical measures that determines the full and uniform germination of seeds. However, as it turns out, in the event of abundant atmospheric precipitation (>70-80 mm) during the seed germination period, post-sowing irrigation may be unnecessary.

The actual water demand of corn and its components under different irrigation regimes and in different years are extracted from the experimental data balance calculation table and are given in the form of an abbreviated Table 3.3.2, in which the actual total demand is expressed by the formula:

$$E = \mu P + m + (\Delta W_1 - \Delta W_2), \qquad (3.3.1)$$

Where W_1 and W_2, respectively, represent the estimated water supply at the beginning and end of the period under consideration:

$\Delta W_1 = r_s - r_{sz}$

$\Delta W_2 = r_b - r_{sz}$ ^

Where r_s and r_b , respectively, are the soil moisture at the beginning and end of the reporting period;

r_{sz} – the required or limiting soil moisture limit – $r_{zR} = 70\%$;

$r_{zR} = 60\%$

The summation of the members included in the balance is done algebraically. The soil moisture reserve with a negative sign shows that there is no moisture reserve in the soil, but rather a deficit compared to the lower limit of the calculated moisture content.

As is known, the water requirements of the plant are ensured by the existence of soil moisture reserves, atmospheric precipitation and, if necessary, irrigation.

In a different case, the water requirement of corn and its components is characterized by the indicators given in Table 3.3.2.

Table 3.3.2

Corn water requirement and its constituent elements

Year	Pre-watering Humidity in % of maximum capacity	Periods	Evaporationჯამური	Atmospheric precipitation	Irrigation with water	Soil moisture content	Total atmospheric precipitation	Usage Coefficient μ
1	2	3	4	5	6	7	8	9
First year	r_{zR} -70% of	„sagvelas aghmotsenebis shemdeg“	259,71	266,70	–	-6,99	266,70	1,00
		%%	100	102,69	0			
		After the emergence of the sagvela"	314,0	54,90	232,07	27,03	66,60	0,82
		%%	100	17,48	73,90			
		Total	579,91	321,60	232,07	20,04	333,30	0,96
		%%	100	55,46	40,03			
First year	r_{zR} -60% of	„sagvelas aghmotsenebis shemdeg“	259,71	266,70	–	-6,99	266,70	1,00
		%%	100	102,69	0			
		After the emergence of the sagvela"	301,47	66,60	210,39	24,48	66,60	1,00
		%%	100	22,09	69,79			

Table 3.3.2 (continued

1	2	3	4	5	6	7	8	9
First year	r_{zR} - 60% of	Total	561,18	333,00	210,39	17,49	333,00	1,00
		%%	100	59,33	37,49			
Third year	r_{zR} -70% of	„sagvelas aghmotsenebis shemdeg"	242,55	161,38	159,16	-79,99	225,50	0,80
		%%	100	66,53	65,62			
		After the emergence of the sagvela"	273,60	80,30	156,16	37,14	80,30	1,00
		%%	100	29,35	57,07			
		Total	516,15	241,68	315,32	-40,85	305,80	0,79
		%%	100	46,82	61,09			
Third year	r_{zR} -60% of	„sagvelas aghmotsenebis shemdeg"	237,57	164,25	82,06	-8,74	225,50	0,728
		%%	100	69,13	34,54			
		After the emergence of the sagvela"	260,21	79,80	205,71	-25,30	79,80	1,00
		%%	100	30,67	79,05			
		Total	497,78	244,05	287,77	-34,04	305,30	0,799
		%%	100	49,03	57,81			
Fourth year	r_{zR} -70% of	„sagvelas aghmotsenebis shemdeg"	310,88	207,27	176,47	-72,86	220,50	0,94
		%%	100	66,67	56,76			
		After the emergence of the sagvela"	162,65	101,47	46,49	-13,30	187,20	0,54
		%%	100	62,38	28,58			
		Total	310,88	207,27	176,47	-72,86	220,50	0,94
		%%	100	66,67	56,76			

Table 3.3.2 (continued)

1	2	3	4	5	6	7	8	9
Fourth year	r_{zR} - 70% of	Total	473,53	308,74	252,96	-88,16	407,50	1,00
		%%	100	65,20	53,42			
Fourth year	r_{zR} -60% of	„sagvelas aghmotsenebis shemdeg"	303,71	207,27	98,42	-2,68	221,20	0,94
		%%	100	68,74	32,40			
		After the emergence of the sagvela"	152,52	128,47	103,49	-79,44	209,40	0,61
		%%	100	84,23	67,85			
		Total	456,23	336,44	201,91	-82,12	430,66	0,78
		%%	100	73,74	44,25			
Fifth year	r_{zR} - of 70%	„sagvelas aghmotsenebis shemdeg"	311,64	204,10	149,42	-41,88	204,10	1,00
		%%	100	65,49	47,94			
		After the emergence of the sagvela"	243,30	75,60	158,40	9,30	75,60	1,00
		%%	100	31,07	65,10			
		Total	554,94	279,70	307,82	-32,58	279,70	1,00
		%%	100	50,40	55,46			
Fifth year	r_{zR} -60% of	„sagvelas aghmotsenebis shemdeg"	304,01	204,90	102,16	-3,05	204,90	1,00
		%%	100	67,39	33,61			
		After the emergence of the sagvela"	221,31	63,60	103,55	54,16	63,66	1,00
		%%	100	28,74	46,79			
		Total	525,32	268,50	205,71	51,11	268,50	1,00
		%%	100	51,11	39,16			

Table 3.3.3

Relative total evaporation by soil moisture before irrigation

period	$E_{0,8}/E_r$	Years					Medium	According to the formula
		First year	Second year	Third year	Fourth year	Fifth year		
1	2	3	4	5	6	7	8	9
Before the appearance of the qochochi	0,8/0,8=1	1	1	1	1	1	1	1
	0,8/0,7=1,143	1	1,017	1,073	1,041	1,050	1,036	1,028
	0,8/0,6=1,333	1	1,037	1,099	1,067	1,074	1,055	1,062
After the appearance of the qochochi	1	1	1	1	1	1	1	1
	1,143	1,027	1,008	1,150	1,052	1,045	1,056	1,057
	1,333	1,070	1,059	1,226	1,157	1,130	1,128	1,028
During the vegetative period	1	1	1	1	1	1	1	1
	1,143	1,015	1,012	1,100	1,046	1,047	1,044	1,041
	1,333	1,037	1,049	1,141	1,110	1,101	1,087	1,090

As expected, water demand reaches a minimum in the driest variants, where irrigation is carried out when the soil moisture content drops to r_{zR} 60% of, while maximum water demand occurs in the wet variants, when the soil moisture content in the pre-irrigation period is r_{zR} 80% of.

As for water requirements in medium-humidity variants, when the soil moisture before irrigation is equal to r_{zR} 70% of., they occupy an intermediate position between the above water requirements.

As can be seen from Table 3.3.3, the total water content under different humidity conditions varies within the following limits:

– When the pre-irrigation humidity is 70% of –573,71÷473,53 mm;

– When the pre-irrigation humidity is 60% of –561,18÷456,2 mm.

Irrigation norms vary within the following limits:

– When the pre-irrigation humidity is 70% of – 396.12-232.07 mm;

– When the pre-irrigation humidity is 60% of – 393,47÷201.91 mm.

The amount of atmospheric precipitation used for the benefit:

– When the pre-irrigation humidity is 70% of – 321,6÷212,45 mm;

– When the pre-irrigation humidity is 70% of – 336,44÷204,28 mm.

The overall analysis of the data shows that the lower the pre-irrigation humidity, the lower the total evaporation. The same can be said about irrigation norms.

The rate of use of natural atmospheric precipitation is quite high, which varies within 343,23÷204,28 mm depending on different irrigation regimes and meteorological conditions r_{zR} 80%of the vegetation period. Under the conditions of the optimal irrigation regime (80%), the share of atmospheric precipitation in satisfying the total water demand is ((6,88÷35,25)%. The highest value – 65.88% – corresponds to the second year, which was distinguished by the amount of precipitation falling in the period from June to September (97,3+ 46,6 +96,2 +91,5 = 331,4 mm), which is more than in all years, and the lowest value corresponds to the third year, when the amount of precipitation falling in the same period (71,1^+^3,5^+^22,3^+^45,6^=^142,5 mm) was the least compared to other years. In the remaining years, the above-mentioned indicator varied within (57,25÷48,00)%of

In a relatively dry year, when the pre-irrigation humidity is r_{zR} 60% of it is necessary to increase the share of atmospheric precipitation in satisfying total evaporation by 2÷8% .

In other years, the rate of atmospheric precipitation varies within(66,61÷48,28)%.

As we can see, the rate of use of atmospheric precipitation in the period before the emergence of the seedling is slightly higher than the rate of use during the entire irrigation period.

The above is also explained by the fact that the region is characterized by a fairly large amount of precipitation in the period from sowing to the emergence of the seedling. As can be seen from Table 3.3.2, the amount of precipitation during this period ranges from 203.7 mm to 266.70 mm in 4 out of 5 years, and once, in the third year, it dropped to 172.7 mm. It should be assumed that the rational use of such a large amount of atmospheric precipitation should lead to significant savings in irrigation water.

Indeed, Table 3.3.2 shows that in one year out of five years (the fourth year), the amount of atmospheric precipitation (266.70 mm) from sowing to the emergence of seedlings was greater than the total water demand. In these years, neither irrigation immediately after sowing nor vegetation irrigation until the emergence of seedlings was carried out. The coefficient of use of atmospheric precipitation was high $\mu = 1$.

In one year out of five years, in the second year, the water deficit was insignificant (21.07 mm). This year, one vegetation irrigation and one irrigation immediately after sowing were carried out. Naturally, it was possible to avoid vegetation irrigation and to irrigate the coffee fields only immediately after sowing, as evidenced by the coefficient of atmospheric precipitation utilization 0,728÷0,80, which dropped to

In three out of five years, the amount of atmospheric precipitation varied from 113.35 mm to 120.76 mm, and in the third year, although irrigation was not carried out immediately after sowing, instead, 3, 2 and 1 vegetation irrigation were carried out according to the options. It is clear that in all irrigation regimes it was possible to limit ourselves to one irrigation each, since the balance allowed for this (113.35 mm). The deficit could be eliminated by conducting two irrigations with a norm of 60÷60 mm.

In the second year, the deficit was in the period before the emergence of the seedling – 291.51 – 172.70 = 118.81 mm. Under the conditions of the high humidity regime, immediately after sowing, two vegetative irrigations were carried out and 205.18 mm of irrigation water was supplied, i.e. 205.18 – 118.81 = 86.37 mm, i.e. much more than the norm for one irrigation. Naturally, in this case, we should have refrained from one vegetative irrigation. This is also evidenced by the relatively low coefficients of atmospheric precipitation use, which in some cases fall to 0,54÷0,61. The fifth year also provides a similar picture to the year under consideration.

Based on all of the above, it should be said that for one year out of five years (the first year), with a 20% guarantee, irrigation was not necessary during the period of seeding. Also, for one year out of five years (the first year), irrigation immediately after sowing was not necessary, but instead it was necessary to carry out two vegetative irrigations. For three years out of five years (1, 3, 4 years), i.e. with a 60% guarantee, it was necessary to carry out one sowing and one vegetative irrigation. Since irrigation immediately after sowing is considered a necessary agrotechnical measure that ensures even and complete seed germination, irrigation immediately after sowing should remain valid, but it may be optional in some exceptional years. (20÷40% With security).

In most cases, one vegetative irrigation is necessary with a 60% coverage. Only in rare cases is it possible to conduct vegetative irrigation with a 20% coverage. It was not necessary at all, or on the contrary, it was necessary to conduct two vegetative irrigations. The expediency of such an irrigation regime is also justified by other considerations, which were indicated by Prof. I.A. Chkhenkeli. The fact is that in spring and early summer (April, May, early June), along with the abundance of atmospheric precipitation, there is often a drop in air temperature, and therefore, conducting irrigations in the period between the emergence of the caterpillars can lead to undesirable consequences. The growth and development of corn is delayed, the vegetative period is prolonged, and ultimately the yield is reduced. Therefore, the first vegetative irrigation must be linked to temperature conditions. In any case, it is preferable not to carry out irrigation in the third decade of June. This will be facilitated by irrigation immediately after sowing, which ensures full and uniform emergence of seedlings, accumulation of moisture reserves for the subsequent period and the possibility of carrying out the first vegetative irrigation later (with sufficient moisture reserves).

The most rational irrigation regime is the regime in which the soil moisture before irrigation is equal to that during the period R_{ZR} of seedling emergence (Table 3.3.2), and in the conditions of the active layer H=o.6 m of the soil, in order to avoid early spring irrigation, the optimal pre-irrigation moisture level should be maintained at r_{zR} 80% in the period after seedling emergence.

Based on the general analysis of Table 3.3.2, it was revealed that the lower the pre-irrigation humidity, the lower the total evaporation. This view is confirmed by the magnitude of the relative total evaporation. $E_{0,8}/E$. Which is given in Table 3.3.3. $E_{0.8}$ means the total evaporation at pre-irrigation humidity when means $r = r_{zR}$ 80%, While E_r is the total evaporation under different pre-irrigation humidity conditions than the above, when r= R_{ZR} 70% end 60%.

.At the beginning of harvesting, at a humidity of 70% of the r_{zr}, on average, 1036 times less water is used(1÷1,073) than at a humidity of 80% of the r_{zr},, and at a humidity of 60% of the r_{zr}, , on average, 1055 times less water is used (1÷1,099).

In the period following the appearance of "qochochi", under the conditions of the r_{zr} 70% humidity regime of the RZR, (1,027÷1,15) times and on average 1.056 times less water is consumed, and under the conditions of the r_{zr} 60% humidity regime of the (1,059÷1,226 times and on average 1.128 times less.

During the entire vegetation period, the following picture occurs on average: under the conditions of the r_{zr} 70% humidity regime of the rzr (1,015÷1,1),, on average – 1.044 times less water is consumed, and under the conditions of the r_{zr} 60% humidity regime of the rzr – (1,037÷1,141) times less, on average 1.087 times less, compared to the r_{zr} 80% humidity regime of the rzr.

As a result of mathematical processing of the data, it was revealed that the following patterns exist:

ა) Before the appearance of the qochochi

$$K = \frac{0,8}{K_h}\left(\frac{80}{r}\right)^{0,21}; \qquad (3.3.2)$$

ბ) After the appearance of the qochochi

$$K = \frac{E_{0,8}}{E_r}\left(\frac{80}{r}\right)^{0,42}; \qquad (3.3.3)$$

გ)On average,vegetacion period

$$K = \frac{E_{0,8}}{E_r}\left(\frac{80}{r}\right)^{0,3}, \qquad (3.3.4)$$

where r– is the pre-irrigation soil moisture from r_{zr} .

The results determined by the formula are given in the same table, from which it is clear that the accuracy of the above formulas is quite acceptable for practice.

If we take these results into account, then we must make an appropriate correction to the calculation formula for the reproducibility - we must divide the right side of the formula by the expression obtained above, or, which is the same, multiply by the reciprocals of these expressions:

$$\frac{1}{K}=\frac{E_r}{E_{0,8}}=\frac{r}{0,8}. \qquad (3.3.5)$$

I.e. formula (3.2.3) will take the following form:

$$E=2,76D^{0,193}\left(\frac{H_2}{0,6}\right)^{1,04}\left(\frac{r}{0,8}\right)^{m}, \qquad (3.3.6)$$

Where the quality indicator is taken:: In the period before the release of "Sagvela" – 0.21, qococis"in the post-emergence period – 0.42 And on average during the entire growing season – 0.30.

From the above, we can conclude that the calculation formula we propose has sufficient accuracy for practice.

3.4. Water requirement of corn and determination of required moisture in the soil

The relative coefficient of water requirements is determined according to the variation of soil moisture in relation to the marginal water holding capacity. Table 3.4.1 gives the values of the relative coefficients of water requirements for different years depending on the marginal moisture.

If we look at the above table, we will notice that the water requirement coefficient decreases as the soil moisture increases. Table 3.4.1 separately shows the values of the ratios of the coefficients of water holding capacity. $K_f / K_a = K_r$,

The relative soil moisture content is approximately equal to 100-80%, 80-70% and 70-60%. For the sake of simplicity of notation, the relative water demand coefficient corresponding to 100-80% of is denoted by 80%. The relative water demand coefficient corresponding to 80-70% is denoted by 70%, and the relative water demand coefficient corresponding to 70-60% is denoted by 60%. In general, the relative water demand coefficient corresponding to a given moisture content is denoted byK_T.

Table 3.4.1

Relative coefficients of water requirements under differen humidity ranges(in% of Kzr.

macveneblebi	Lower limit of soil moisture		
	100-80	80-70	70-60
1	2	3	4
First year	1,126	1,021	0,995
	1,125	1,054	0,997
	1,053	0,988	
$\overline{K}_t$ Medium	1,101	1,024	0,966
$\overline{K}_r / K_{80}$ Ratio of averages	1,000	0,930	0,877
Second year	1,128	0,931	0,874
	1,045	1,098	0,868
	1,075	0,988	
$\overline{K}_t$ Medium	1,082	1,044	0,871
$\overline{K}_r / K_{80}$ Ratio of averages	1,000	0,964	0,806
Third year	1,018	0,846	0,765
			0,764
$\overline{K}_t$ Medium	1,018	0,846	0,759
$\overline{K}_r / K_{80}$ Ratio of averages	1,000	0,831	0,745
Fourth year	1,000	0,965	0,84,^
	0,982	0,990	0,968
	1,090	0,967	0,965
			0,845
$\overline{K}_t$ Medium	1,024	0,974	0,875
$\overline{K}_r / K_{80}$ Ratio of averages	1,000	0,951	0,845

Table 3.4.1 (continued)

1	2	3	4
Fifth year	0,998	0,918	0,922
	1,078	0,898	0,901
$\overline{K}_t$ Medium	1,003	0,913	0,907
$\overline{K}_r / K_{80}$ Ratio of averages	1,000	0,910	0,904
$\overline{K}_t$ Multi-year average	1,0456	0,9602	0,8756
$\overline{K}_r / K_{80}$ Multi-year average ratio	0,918	0,917	0,837

Analysis of Table 3.4.1 shows that when soil moisture varies within r_{zR} 100-80% of the range, the deviation from the water demand calculation formula in individual periods varies within 0.2÷12.8%, and during the entire vegetation period it varies from 1% (second year) to 10%. The average deviation according to multi-year data is 4.5%.

As the soil moisture range decreases, the water demand coefficient also decreases in a linear manner. If we consider the water demand coefficient equal to one when the soil moisture varies within r_{zR} 100-80% of , then the water demand coefficient under different moisture conditions is: 83÷96%, when the moisture is within r_{zR} 80-70% of ; 74.5÷90.7%, when the soil moisture varies within r_{zR} 70÷60% of .

According to the average multi-year data, the water demand coefficient is 91.8%.

As a result of mathematical processing of the data, it was found that the correlation equation between relative evaporation and soil moisture has the following form:

$$\frac{E_r}{E_{80}} = \frac{K_r}{K_{80}} = 0{,}32 + 0{,}85r\,, \qquad (3.4.1)$$

Where E_{80} and K_{80} is the water demand and the water demand coefficient when the soil moisture content varies within the range of r_{zR} 100÷80%. F_r And K_r the same when the soil moisture content differs from the above and expresses the lower limit of soil moisture in the soil. I.e. The values should be inserted into the formula: 0.8, 0.7 and 0.6.

If we take into account the above formula, then the formula for the water demand coefficient can be presented as follows:

$$K_s = K_r\left(\frac{E_r}{E_{80}}\right) = K_r(0{,}32 + 0{,}85r). \qquad (3.4.2)$$

Water demand is expressed by the following formula:

$$E = \left[0{,}6\left(\frac{D}{6{,}7}\right)^{0{,}807}\right]\left(\frac{H}{0{,}6}\right)^{1{,}04} D = 2{,}76D^{0{,}193}(0{,}32 + 0{,}85r). \; (3.4.3)$$

As we can see, the last formula differs from formulas (3.4.1) and (3.4.2). The point is that formula (3.4.2) is used when the pre-irrigation moisture content is measured, while formula (3.4.3) is used when the soil moisture content changes within a certain range to r_{zR} 100-80%, 80-70% and 70-60%. In the case of the previous formula, when, for example, the pre-irrigation moisture content is 70-60%, the water demand will go through three phases – 100-80%, 70-70% and 70-60%, and therefore, the total water demand will reflect the influence of all three phases.

If we want to use the last obtained formula and the final data of its corresponding table in the case when the pre-irrigation soil moisture is equal to r_{zR} 70% of , then the water demand, which occurs under the conditions of pre-irrigation moisture of r_{zR} 80% of , should be multiplied (according to the data of the last line of Table 3.4.1) by the coefficient 2*1 + 0.918 = 0.971, and if we want to use it in the case when the pre-irrigation moisture is r_{zR} 60% of , then it should be multiplied by the coefficient 2*1 + 0.918 + 0.837 = 0.938. I.e. In the first case, the water demand decreases by 3%, and in the second case by approximately 6%.

The total water demand coefficient represents the total water demand per unit of crop, i.e. the amount of water consumed per unit of crop (1 quintal, 1 ton) in m^3. Its unit is m^3/t or m^3/t.

Table 3.4.2 presents the total water demand coefficients determined on the basis of experimental data. As can be seen from the table, the lower the total water demand coefficient, the better the irrigation regime and the higher the mineral fertilizer application rates. In all cases, the total water demand coefficient decreases as the yield increases. It has the smallest values when the pre-irrigation moisture content is equal to r_{zR} - 80% of and the norms of irrigation and with it the yield are maximum. In this case, the total water demand coefficient varies within 52.337÷0.79 m^3/c and on average over five years amounts to 61.32 m^3/c. The total water demand coefficient takes on the greatest value when, in non-fertilized variants, the preirrigation soil moisture content is equal to 60%. In this case, the water demand coefficient varies within 189.3÷306.32 m^3/c, and on average amounts to 244.79 m^3/c.

The above-mentioned correlations between total water demand and yield, under any irrigation regime, indicate that the total water demand coefficient should be expressed by the kinetic production function of E. Hede and D. Dillon, as well as yield. For this, the total water demand coefficient K should be written in the yield matrix.

Otherwise, the matrix will remain unchanged. Having analytical formulas for the K coefficient and yield, we can determine the total water demand E=Ky for any irrigation regime, precipitation norms, and meteorological conditions.

Table 3.4.2

The impact of different irrigation regimes and mineral fertilizer rates on the total water demand coefficient

ტენიანობა, % ნია-	Fertilizer norms	First year			Second year			Third year			Fourth year			Fifth year			5-year average		
		Total Water demand Value, mm	Grain yield, c/ha coefficient, cbm/s	კოეფიციენტი, კბმ/ც Totsl Water-soil coefficient, cbm/s	Total Water demand Value, mm	Grain yield, c/ha	Totsl Water-soil coefficient, cbm/s	Total Water demand Value, mm	მარცვლის მოსავალი, ც/ჰა	Totsl Water-soil coefficient, cbm/s	Total Water demand Value, mm	Grain yield, c/ha	Totsl Water-soil coefficient, cbm/s	Total Water demand Value, mm	Grain yield, c/ha	Totsl Water-soil coefficient, cbm/s	Total Water demand Value, mm	Grain yield, c/ha	Totsl Water-soil coefficient, cbm/s
1	2	3	4	5	6	7	8	9	10	11	12	13	14	15	16	17	18	19	20
70%	Fertilizer-free	573,71	32,20	178,17	516,15	36,05	143,17	473,53	37,50	126,27	554,94	30,50	181,94	557,23	36,75	152,62	535,11	35,00	152,88
	$N_{90}P_{75}K_{60}$		51,20	112,05		51,94	99,37		63,09	76,05		52,11	106,49		56,34	98,90		54,93	97,41
	$N_{120}P_{125}K_{90}$		62,84	91,29		63,50	81,28		72,42	65,11		59,52	93,23		41,42	78,02		66,00	81,07
	$N_{200}P_{175}K_{120}$		68,44	83,82		71,60	72,08		77,88	60,89		64,68	85,79		76,18	73,14		71,75	74,57
60%	*Fertilizer-free*	561,18	18,32	306,32	497,78	20,63	241,48	456,23	24,10	189,30	525,32	19,90	263,97	530,02	22,05	240,37	514,10	21,00	244,76
	$N_{90}P_{75}K_{60}$		35,56	157,81		36,04	138,11		40,22	113,10		33,22	158,13		40,37	131,29		37,08	138,61
	$N_{120}P_{125}K_{90}$		42,68	161,48		44,00	113,13		52,96	86,14		43,76	120,04		46,66	113,35		46,01	111,73
	$N_{200}P_{175}K_{120}$		45,53	123,25		49,60	100,35		56,48	80,77		45,56	112,82		53,00	100,00		50,20	102,39

Thus, a clear correlation has been established between meteorological factors and soil moisture, which in itself determines the determination of total water requirements. The latter is integrally reflected in the quantitative indicators of the harvest.

Chapter IV.
Conditions for conducting field experiments

4.1 Predictable yield of corn and methodology of agrotechnical measures

The experiment was conducted on the territory of the Mukhrani Experimental Farm of the Georgian State Agrarian University in a five-year cycle, according to the following scheme:Soil

Soil R_{zr} 60-**55%**

1. No fertilizer (control)
2. $N_{90}P_{75}\,K_{60}$
3. $N_{120}P_{125}\,K_{90}$
4. $N_{200}P_{175}\,K_{120}$

Soil R_{zr}((70-65%)

5. No fertilizer (control)
6. $N_{90}P_{75}\,K_{60}$
7. $N_{120}P_{125}\,K_{90}$
8. $N_{200}P_{175}\,K_{120}$

The experiment used the corn variety `Ajameti Tetri'. The experimental scheme consisted of 8 variants. The experiment used a plot area of 10 m * 10 m or 100 m^2, 4-fold replication, 3-fold replication in time. In order to exclude the influence of moisture and fertilizers from neighboring variants, the calculated areas were delimited by 5-meter-wide protective strips. Phosphorus and potassium fertilizers were applied before tillage, in accordance with the norms provided for in the experimental scheme. One mesa-med of nitrogen fertilizers was applied before pre-sowing cultivation, and the rest was applied as top dressing in the 6-7 and 10-12 leaf phases. From mineral fertilizers, superphosphate (20% MN), potassium salt (40% MN), ammonium sulfate before pre-sowing cultivation, and ammonium nitrate (34% MN) were used during feeding.

We sowed with 0.70 m row spacing, with a planting density of 57 thousand plants per hectare. Agronomic techniques included four-fold cultivation of the rows, application of mineral fertilizers and loosening.

In order to determine the optimal irrigation dates, we systematically studied the dynamics of moisture in the active layer.

During the field trials, the following observations, determinations and records were made:

- Phenological observations of corn growth and development were made in the 5-6 leaf, 10-11 leaf, cob emergence, flowering, milky and waxy maturity phases of germination;
- Soil moisture content in the active layer;
- Number of irrigations, irrigation and irrigation rates, total water demand during the growing season;
- Determination of plant height and assimilation leaf surface area;
- Differential recording of the intensity of formation of fresh and dry biomass;
- Photosynthetic potential;
- Net productivity of photosynthesis in crops;
- Recording of actinometric indicators;
- Determination of nutrients in the soil: nitrate and ammonia nitrogen, mobile phosphorus and exchangeable potassium throughout the growing season;
- Plant establishment frequency;
- Recording of green mass yield;
- Determination of the chemical composition of green mass, energy food units and digestible protein;
- Extraction of nutrients;
- Processing of experimental results by the method of dispersion analysis.

The humus content in the soil was determined by the Tyurin method, easily soluble nitrogen by I.U. Konovalov, total phosphorus by Denizh, nitrate nitrogen reagent and ammonia nitrogen by Konev, mobile phosphorus by Kirsanov, exchangeable potassium by Puseinika and Protamosov.

To determine crude protein, nitrogen content is multiplied by a coefficient, the numerical value of which is taken equal to 6.25. Nitrogen-free extractable substances are determined by various methods, and the nutritional value of green mass is determined using the conversion method.

We conducted experimental statistical data analysis in the Laboratory of Electronic Computing Machines of the State Agrarian University, using dispersion and regression methods.

To determine the assimilation areas of the leaf surface, we use the calculation method for twenty labeled plants in each of the four replicates, using the following formula:

$$S = k \cdot a \cdot b \qquad (4.1.1)$$

Where K – the correction coefficient takes into account the deviation of the study area from the equivalent area of a regular geometric shape and in our case is taken equal to 0.67;

a and b – are the average statistical indicators of leaf length and width, respectively.

The total height of the variety used in our experiments, `Ajameti Tet-ri', is 2-2.5 m, the stem is weak, the number of leaves is 17-20, the mass of 1000 grains is 0.340.40 kg, the vegetati on period from germination to full maturity is 140-155 days.

The aim of the research is to scientifically study and substantiate the optimal parameters of artificially regulated crop-producing factors - soil cultivation and fertilizer application system, irrigation techniques and regime, crop protection from weeds and pests, production organization, and others, taking into account soil-climatic and biopotential productivity.

Therefore, the research conducted by us included the study of the following issues:

- Specific water-physical properties of the soil;
- Natural agro-climatic resources: radiation balance, thermal regime, atmospheric precipitation and the nature of its distribution during the vegetation period;
- The influence of various irrigation regimes and mineral fertilizer doses on plant growth and development;
- Determination of calculation parameters of meteorological conditions corresponding to the provision of a given region according to statistical-variational data in a multi-year rate;
- Chemical composition of corn grain and straw in the case of different irrigation regimes and different doses of mineral fertilizers;
- Determination of the coefficient of use of mineral fertilizers according to the output of nutrient elements;
- Determination of the optimal irrigation regime taking into account the water requirements of corn, variability of meteorological factors and soil moisture before irrigation.

The prospect of increasing grain yields, and in particular corn yields, determines the transition to a new system of production planning and management based on mathematical methods and the widespread use of EGM.

The advantage of this system lies in the complex accounting and timely assessment of the factors determining the processes.

The introduction of the principle of systematic analysis in the production of grain corn allows us to maximally take into account the influence of both the biological characteristics of the plant and a wide range of operating soil-climatic, agrotechnical and organizational factors on the yield. From this point of view, programming implies the use of a scientific method to select optimal planning and technological measures to obtain the highest possible yield on a separate field or on the entire plot.

Mathematical probabilistic-statistical methods are widely used in crop programming and forecasting.

The dependence of the yield of a crop with a certain genetic biopotential on environmental factors: radiation, thermal regime, soil moisture dynamics, nutritional elements, etc. It is quite well reflected by exponential dependences. Therefore, our research is based on the use of existing traditional and new progressive methods of comparative research

In our research, great importance is attached to the study of specific hydrophysical properties of soil and the simultaneous determination of relevant parameters. In this regard, some attention has been paid to the study of the specific surface area of soil as a hydrophilic capillary-porous physical body.

It is known that when water passes from a free state to a phase state, it is expressed in the absorption of water by the surfaces of soil aggregates, which is accompanied by the release of energy and the release of heat. Often, this latter leads to the breakdown of the soil structure, its dispersion. It is the specific heat of wetting, which in a certain range of humidity is linearly related to the density of the surface, that determines the degree of dispersion in soils with both immobile and mobile lattice structures, and the latter mainly depends on the mineralogical composition of the soil particles.

Analysis of special laboratory studies showed that the soils of the study site (Mukhran) are characterized by quite high water resistance and undergo insignificant evaporation, which is mainly due to the presence of clay-colloidal particles containing potassium.

Therefore, the water-air regime in the active layer of the soil (Mukhran) does not require regulation by taking special measures.

Based on the probabilistic-statistical processing of meteorological data on a multi-year basis, a new criterion was adopted for assessing the total impact of environmental and regional water supply factors from a climatic and meteorological point of view.

In the specified soil-climatic and agro-ameliorative zone of Shida Kartli, the possibility of optimal management of the corn variety - `Ajameti Tetris' - has been determined based on the identification of one of the correlation links between water demand and mineral fertilizers.

The main results of our scientific research work were reflected in the scientific thematic plan of the Department of Hydrotechnics and Melioration of the Agricultural Institute. ^

For the characteristic regions of Georgia, in particular, Shida Kartli, the optimal norms for the use of mineral fertilizers have been determined based on the water demand of the plant and the establishment of appropriate irrigation regimes.

The yield of the corn variety ``Ajameti Tetri" is regulated by taking into account the maximum possible number of statistical and climatic variables and it has been established that it is possible to obtain it at the level of 50-80 t/ha.

The methodology proposed by us for the cultivation of programmed corn crops and the implementation of relevant technologies in practice ensure a high and stable yield.

The new, specified criteria for assessing the region's water supply and the proven method for determining the water requirements of plants can be widely used both in similar and different meteorological conditions to determine the irrigation regime and water consumption of agricultural grain crops.

Chapter V.
Results and analysis of corn yield prediction

5.1. Results of phenological observations

In accordance with the existing methodology, in order to determine the influence of different irrigation regimes and mineral fertilizer rates on plant growth and development, phenological observations were planned to be conducted throughout the entire period of the experiment.

In all variants of the experiment, the dates were recorded. Sowing, germination, leaf phase, ear emergence, whisker emergence, flowering, maturity - milky, waxy, full. The main results of these observations are given in Table 5.1.1.

Table 5.1.1

Phenological observations on the development of the maize variety \`Ajameti Tetri' in a five-year cycle

Option #	Sowing time	germinathe beginning of corn		2nd and 3rd leaf phase	Qococ გამოტანა		Mustache გამოტანა		Flowering		Maturity		
		Initial	complete		Initial	complete	Initial	complete	Initial	complete	Milky რძისებრი	Wax-sebrium	complete
1	2	3	4	5	6	7	8	9	10	11	12	13	14
Soil R_{zR} 60-55%													
1	18,04	8,05	12,05	16,05	8,07	15,07	22,07	24,08	26,08	30,08	3,09	6,09	15,09
2	18,04	8,05	12,05	16,05	8,07	15,07	22,07	24,08	26,08	30,08	3,09	6,09	15,09
3	18,04	8,05	12,05	16,05	8,07	15,07	22,07	28,08	26,08	31,08	3,09	6,09	15,09
4	18,04	10,05	12,05	16,05	10,07	15,07	24,07	28,08	29,08	31,08	3,09	6,09	15,09
Soil R_{zR} 75-70 %													
5	18,04	8,05	11,05	15,05	8,07	16,07	23,07	26,08	28,08	30,08	2,09	6,09	16,09
6	18,04	8,05	11,05	15,05	8,07	16,07	23,07	26,08	28,08	30,08	3,09	6,09	16,09
7	18,04	8,05	11,05	15,05	8,07	16,07	23,07	26,08	28,08	30,08	3,09	8,09	18,09
8	18,04	10,05	11,05	19,05	10,07	18,07	24,07	28,08	29,08	31,08	2,09	8,09	18,09

Phenological observations showed that the corn variety \`\`Ajameti Tetri" sprouts appear on the 20th day under the soil-climatic conditions of the zone, and full germination occurs on the 22nd-25th day after sowing. Similar results were obtained in lysimeters for all plots of the experiment.

The second-third leaf phase began almost simultaneously, with a slight difference (2-3 days) depending on the plots.

Figure 1. The best option

The fourth-fifth leaf phase began with a difference (3-4 days) depending on the divisions.

Figure 2. The best option

The growth rate of corn plants in height varied significantly during the optogenesis period depending on environmental conditions. In the first 15 days after emergence, the average daily growth was 9.05-12.05 cm, and in the next two to three weeks, it was quite intense (1.6-2.5 cm), and in the next two to three weeks, it was 0.4 cm per day.

Figure 3. The best option

The further growth rate gradually increased and reached a maximum on the 7th-12th day. The cob emergence stage occurred almost simultaneously in all plots. The difference was only 2 days, and the end of this phase was 2-3 days earlier in variants 6, 7 and 8, while for the flowering phase this difference was significantly reduced (Table 5.1.1).

Maize grain maturity was observed in three sub-phases: milky, waxy and full maturity. It, like flowering, began on average 1 day earlier in the last four variants. Waxy maturity 4 days earlier than in the control. This is explained by the increase in nutrient rates and soil carrying capacity.

Figure 4. The best option

The full maturity of corn grains was reached 4-5 days earlier in the better variants, depending on the yield.

Summarizing the results of phenological observations, we can note that no sharp differences were observed in the development of corn depending on the variants.

Figure 5. The best option

It should be noted that germination occurred on the 21 st and 23rd day after sowing, tillering and moustache emergence on the 88th and 102nd days after sowing, flowering on the 108 th day, and full maturity on the 134th day. The time from flowering to full maturity was 28 days.

5.2. Results of biometric analysis

The results of biometric analysis and measurements show that the structural elements of the corn crop differ significantly depending on the variants (Table 5.1.1).

Under conditions of 60-55% of the soil's marginal capacity, there are a total of 350 plants on the no-fertilizer variant (1). The average weight of one plant is 0.44 kg. The number of leaves per plant is 16.2. There are a total of 338 plants with taro and 12 non-taro plants, 13 with two-taro, 325 with one-taro, the average height of the plant is 1.64 m. In the fourth variant, the number of plants is 402, a total of 398 plants with taro, 4 plants without taro, 18 plants with two-taro, 380 plants with one-taro. The average weight of one plant is 0.64 kg, the average number of leaves is 18.2, and the average height of the stem is 1.74 m.

Under the conditions of 70-65% of the soil R_{zR} marginal capacity, there are 355 plants on the control unfertilized variant (5). A total of 331 plants are taro. Among them, 310 are single-stemmed, 21 are double-stemmed. There are 24 plants without taro. The average weight of one plant is 0.88 kg, the average number of leaves is 17.6, and the average height of the stem is 1.89 m. The best, eighth variant, has 391 plants. A total of 380 plants are taro. There are 11 plants without taro. There are 368 plants with single-stemmed, and 12 plants with double-stemmed. The average height of the stem is 2.70 m. The average weight of one plant is 0.87 kg, and the average number of leaves per plant is 18.

Despite the same irrigation and nutrition conditions, the average height of corn increases sharply. The structural elements of the corn crop testify to its potential capabilities in the given ecological conditions when using appropriate agricultural techniques. It should be noted that in the variants relatively provided with irrigation and nutrition elements, the number of double-stemmed plants increases by 2-3%, while the number of undouble-stemmed plants decreases in these variants by 0.3-1.2%. The increase in the average plant weight was also maximally observed. Also, foliage was significantly observed in the better 4th and 8th variants of the experiments, by an average of 14% and 12%, respectively, which leads to an increase in both grain and straw yields

5.3. Corn grain and straw yield

The accumulation of dry matter by corn proceeds slowly throughout the vegetation, then increases. Three stages of dry matter accumulation are distinguished: the first - from germination to the 7-8 leaf phase, the second - from the 7-8 leaf phase to the beginning of flowering, and the third - fruiting. All plants accumulate 5% of their total dry matter at the first stage of crop formation, at the end of vegetation. At the second stage, depending on the variety, 27-48% accumulates, and at the third stage - 49-68% of dry matter (Liparteliani, 1985).

The assimilated surface of the plant is the basis for the synthesis of light energy and organic matter, A.A. Nichiporovich, studying the dependence of light energy absorption on the area of corn leaves, came to the conclusion that an area of 45-50 thousand m3/ha with a sufficient amount of moisture and mineral substances can give the greatest yield.

In our studies, light, heat, humidity and the level of nutrition had a certain influence on the intensity of organic matter accumulation in the yield of the corn variety ``Ajametis Tetris".

The accumulation of dry mass by the plant during the growth process occurred throughout the entire vegetation period, and with the appearance of 11-13 leaves, the rate of accumulation of dry matter per unit area of the plant continued relatively slowly: during this period, corn accumulated, on average, 14-18% of the total mass by the end of the vegetation period. At the beginning of vegetation, the mass of dry matter increased mainly due to the growth of leaves, then the mass of the stem gradually increased, and relatively intensive accumulation of organic mass occurred during the period of pod production.

The nutritional conditions of the plant had the greatest impact on the accumulation of dry matter. Different provision of the plant with essential elements and the marginal soil capacity in the experiment had a negative impact on the uniform development of corn in all variants, delayed the emergence of leaves and extended the interphase period in some of them.

In our studies, the influence of the nutrition and irrigation regime on the yield of the `Ajameti Tetri` variety of corn played a major role. At 60-55% of the soil's maximum capacity in variants 4 and 8, the average grain yield over five years was 5.09 and 7.10 t/ha, while at 70-65% of the soil's maximum capacity in variants 6 and 8, the average yield over five years was 7.80 and 8.90 t/ha.

The experiments showed a fairly pronounced positive effect of each type of fertilizer on the accumulation of dry matter by corn.

According to the research data, which are given in Table 5.3.1, the maximum grain yield of 5.65-8.97 t/ha was obtained in variants 4 and 8, where mineral fertilizers N1200P175 K120 were applied and the soil's marginal capacity was 60-55%. It is worth noting that in the control variant without fertilizers, the average yield over five years at 60-55% of the marginal capacity of the soil was 3.70 t/ha, and at 70-65% of the marginal capacity - 7.10 t/ha.

Table 5.3.1

Corn grain yield over a five-year period (t/ha)

Option #	Years of the experiment"					On average 5 years apart
	1	2	3	4	5	
1	2	3	4	5	6	7
Soil R_{zR} 60-55 %						
1	3,55	3,60	4,02	3,32	4,03	3,70
2	4,26	4,40	5,29	4,37	4,66	4,61
3	4,55	4,96	5,64	4,65	5,30	5,09
4	4,76	5,07	5,65	4,63	5,42	5,10
Soil R_{zR} 70-65%						
5	5,12	3,60	6,30	5,21	5,36	5,11
6	6,28	5,19	7,27	5,95	7,14	6,35
7	6,84	6,35	7,78	6,46	7,61	7,80
8	7,73	7,16	8,97	7,07	8,47	8,90
Uas 0,5	1,41	0,83	1,38	1,63	0,72	0,88

Thus, our studies have shown that different doses of mineral fertilizers provide an increase in corn yield. With increasing doses of mineral fertilizers, corn productivity increases, which was ensured by increasing the assimilation surface of the leaves. In all variants, including the control, the yield of corn grain and straw was quite high, which was facilitated by the biological nitrogen accumulated by the predecessors - soybeans and alfalfa.

During the five-year period of the experiment, the yield of corn straw was on average higher (3.90 t/ha) over five years at 60-55% of the marginal soil fertility in variant 4, where mineral fertilizers $N_{200}P_{175}$ K_{120} were applied. Straw yield is also high in option 8, where mineral fertilizers $N_{200}P_{175}$ $K_{120}0$ were applied and the soil's maximum water holding capacity was 70-65% (Table 5.3.2).

Table 5.3.2

Corn straw yield over a five-year period (t/ha)

Option #	Years of the experiment"					On average 5 years apart
	1	2	3	4	5	
1	2	3	4	5	6	7
Soil R_{zR} 60-55 % (ნიადაგის ზღვრული ტენტევადობის 60-55%						
1	2,73	2,76	3,00	2,55	3,10	2,85
2	3,27	3,38	4,06	3,36	3,58	3,55
3	3,50	2,69	4,33	3,57	4,07	3,65
4	3,66	3,90	5,35	3,56	4,16	3,90
Soil R_{zR} 70-65% (ნიადაგის ზღვრული ტენტევადობის 70-65%)						
5	3,93	2,76	4,84	4,00	4,12	3,95
6	4,83	4,00	5,59	4,57	5,49	4,90
7	5,26	4,80	6,00	4,96	5,85	5,40
8	5,94	6,50	6,90	6,43	6,51	7,10
Uas $_{0,5}$	0,43	0,87	1,03	0,39	1,02	0,33

The results of the conducted studies indicate that the adopted agrotechnical measures in the given nutrient rotation ring ensure an increase in the assimilated surface of the plant and thus contribute to the intensive accumulation of biomass.

Against the background of biological nitrogen, post-harvest growth is promoted by all types of fertilizers used according to the experimental scheme. Of all the elements of mineral nutrition for corn in the given soil-humidity-climatic conditions, this type of nitrogen is included in the first minimum and the effect of using mineral fertilizers at the rate of N90P75K60 gives an increase in yield by only 9-13%.

The data presented in Tables 5.3.1 and 5.3.2 show that increasing doses of irrigation and mineral fertilizers ensure an increase in grain and straw yields every year. That is, there is a growth trend.

5.4. The influence of irrigation and fertilizers on corn yield

Regulation of irrigation and feeding regimes taking into account the biological characteristics of the corn plant, the course of phenological phases and the intensity of growth and development give the following picture: from sowing to emergence there are no noticeable differences between the variants.

In the crops of the corn variety ``Ajameti Tetri", the tillering phase and subsequent phases proceeded faster and more intensively (5-15 days) when irrigation was carried out when the soil moisture content was 70-65% of the maximum moisture capacity according to the cartogram $N_{120}P_{175}K_{120}$. It is noteworthy that in conditions of poor fertility and low pre-irrigation humidity, this process proceeded with very low intensity. With the improvement of the nutrient irrigation regime, the tillering process reaches strong and maximum indicators in the variants where the pre-irrigation moisture content is 70-65% of the maximum holding capacity, and the fertilizer rates are maximum.

Data illustrating the influence of the irrigation regime and mineral fertilizer doses on the structure and yield of the green mass of the corn variety ``Ajameti Tetri" are given in Table 5.4.1.

While the variation of the irrigation and nutrient regime did not have a significant effect on the course of the growth and development phases, the impact of these factors on the intensity of tillering was clearly evident. In the 8th variant, the number of shoots increases in accordance with the pre-irrigation soil moisture and the fertilizer rates. However, it is worth noting that increasing fertilizer rates led to an increase in the number of shoots before the application of $N_{120}P_{175}K_{120}$, and the subsequent increase in most cases led to a decrease in these indicators. For example, in the unfertilized variant, respectively, at 70-65% pre-irrigation moisture, the number of shoots increased from 0.8 to 1.0, while increasing the mineral fertilizer rates to $N_{200}P_{175}K_{120}$ reduced the number of shoots, respectively, to 1.0, 1.2 and 1.3.

The experimental analysis shows that the corn variety ``Ajameti Tetrishi" produced a greater number of tillers when we carried out irrigation at a soil moisture content of 70-65% of the maximum allowable moisture content and applied mineral fertilizers at a rate of $N_{120}P_{125}K_{90}$.

The positive effect of the improvement of irrigation and mineral fertilizer regimes on the formation of tiller stems of the plant can be explained by the fact that they ensure an increase in moisture and nutrient reserves in the soil, their relatively more complete absorption by the plant, and an increase in the intensity of cell division and reproduction. This process, in turn, reduced the natural selection of tiller stems, maximally supplying them with nutrients and water.

Table 5.4.1

Impact of irrigation and fertilizers on corn green mass structure and yield

Option #	Fertilizer rates, kg/ha Sodium, phosphorus, potassium	Ten Capacity	corn kernel number / quantity	Plant height	Number of leaves	Leaf area	Mass of one plant, g	Silage green mass yield, t/ha
1	2	3	4	5	6	7	8	9
1	No fertilizer (co-ntrol)	60-55%	0,8	1,64	16	43,4	296,4	10,34
2	$N_{90}P_{75}K_{60}$	60-55%	1,0	1,81	17	47,6	399,3	19,78
3	$N_{120}P_{125}K_{90}$	60-55%	1,1	1,86	18	28,1	498,5	35,22
4	$N_{200}P_{175}K_{120}$	60-55%	1,2	1,96	18	49,0	545,8	37,02
5	No fertilizer (co-ntrol)	70-65%	0,9	1,89	17	46,2	363,7	25,03
6	$N_{90}P_{75}K_{60}$	70-65%	1,1	2,58	18	49,0	556,8	38,41
7	$N_{120}P_{125}K_{90}$	70-65%	1,3	2,69	18	52,6	520,8	43,66
8	$N_{200}P_{175}K_{120}$	70-65%	1,2	2,70	19	51,4	618,9	43,44

Improving the pre-irrigation moisture and mineral nutrition regime promoted not only the formation of suckers on the main stem of the plant, but also its growth. For example, under conditions of pre-irrigation moisture at 60-55% of the maximum yield, the height of the corn variety ``Ajameti Tetri" under the unfertilized variant doses was 164.1 cm. Under conditions of pre-irrigation moisture at 70-65% of the maximum yield, this indicator was 189 cm. As in the previous case (formation of suckers), a further increase in the norms of mineral fertilizers had a negative effect on the height of the plants. As studies by a number of scientists have shown, mineral fertilizers, especially nitrogen fertilizers, contribute to the accumulation of nitrogen in chlorophyll grains, and high humidity, in turn, enhances the transport of nutrients from the roots to the leaves, and, accordingly, the intensity of photosynthesis.

The formation of leaves in plants depends significantly on both their biologica and genetic properties and environmental factors, among which the provision of water and nutrients is considered one of the main factors.

Analysis of the table clearly confirms what we have said above. In particular, the smallest leaves (16-19) are characteristic of plants in unfertilized variants and in conditions of the lowest pre-irrigation humidity. The increase in fertilizer rates and lower pre-irrigation moisture content slightly, but still ensured an increase in the average number of leaves on the main stem of one plant, the number of leaves on the bare stems was within 6-10. In addition, the minimum number of leaves was recorded in the variants where we carried out irrigation at a soil moisture content of

60-55% of the maximum holding capacity and did not apply mineral fertilizers. The maximum was recorded in the conditions of a soil moisture content of 70-65% of the maximum holding capacity and the application of $N_{120}P_{125}K_{90}$ It is noteworthy that leaf development was also better in these variants.

The main components of the pigment system of corn, as well as other agricultural crops, are chlorophyll and carotenoids, which are concentrated in the most important structural elements of the leaf cell - chloroplasts. As early as 1887, Timiryazev wrote: `...the leaf creates the most important food necessary for the plant in qualitative and quantitative terms, it can be said that the life of the plant itself is expressed in the life of the leaf. The plant is the leaf.'

Among the factors that determine the efficiency of the plant's use of solar energy, the leaf surface, its structure, and the physiological state of the leaves in general play an immeasurably large role.

Our research in this direction has shown that mineral fertilizers significantly increase the surface area of the leaves, and improving the irrigation regime significantly increases their effectiveness. In the best options, where we carried out irrigation at 70-65% of the maximum moisture content when calculating the moisture content and applied mineral fertilizers at the rate of N120P125K90, the leaf area per plant reached its maximum value (52.6 dm2).

Irrigation feeding regimes, influencing the formation of shoots, their growth and development, the number and surface area of leaves on the plant, significantly determine the mass of one plant, which, in the case of equal planting frequency, is the most complete determinant of the yield of green mass. Under unfertilized conditions of low pre-irrigation humidity (60-55%), the mass of one plant of the corn variety ``Ajameti Tetris" was 296.4 g. Accordingly, with the improvement of irrigation and feeding regime, this indicator increased depending on the variants and reached a maximum (545.8 g, and 618.9 g).

As already mentioned, under conditions of equal standing moisture, the mass of one plant, which is a function of the plant's stem, height, number of leaves and their surface area, also determines the yield index. Thus, it is clear that factors positively affecting them also ensure the formation of the maximum yield of green mass. For example, in the crops of the corn variety ``Ajametis Tetris", the normal rate of green mass yield of 10.34 t/ha was formed in variants where irrigation was carried out when the soil moisture content was 60-55% of the maximum holding capacity and we did not conduct split-feeding. Accordingly, by improving the irrigation and feeding regime of the corn variety ``Ajameti Tetris", the yield also improved and it became 37.0 t/ha, and when we carried out irrigation at 70-65% of the maximum soil moisture capacity, the yield in the unfertilized variant was 25.0 t/ha, and in the case of N120P125K90 mineral fertilizers, it was 43.6 t/ha.

In agricultural conditions, such an effective means of crop growth as nitrogen fertilizers, in some cases leads to anthropogenic spoilage of products, especially such a phenomenon as the accumulation of nitrates and nitrites. These substances have a toxic and carcinogenic effect on humans and animals.

Nitrates in excess of the permissible norm are converted into nitrites in the digestive tract, which reduces the blood's ability to carry oxygen. That is why their maximum permissible concentration (MPC) has been established in food products.

According to studies, the MPC in one kg of green mass can be equal to 300-400 mg.

Table 5.4.2

The impact of pre-irrigation moisture and mineral fertilizer rates on corn yield (t/ha)

#	Option	Repetitions				
		I	II	III	IV	Medium
1	2	3	4	5	6	7
Soil R_{zR} 60-55 %						
1	No fertilizer (co-ntrol)	19,62	20,65	20,03	21,06	20,34
2	$N_{90}P_{75}K_{60}$	30,40	29,41	39,29	39,02	29,78
3	$N_{120}P_{125}K_{90}$	65,15	34,20	35,99	35,54	35,22
4	$N_{200}P_{175}K_{120}$	36,57	37,85	37,43	36,31	37,04
Soil R_{zR} 70-65%						
5	No fertilizer (co-ntrol)	25,39	24,20	25,69	24,84	25,03
6	$N_{90}P_{75}K_{60}$	39,14	37,82	37,60	39,08	38,41
7	$N_{120}P_{125}K_{90}$	23,04	43,09	44,07	43,65	43,63
8	$N_{200}P_{175}K_{120}$	43,20	42,90	43,95	43,80	43,44

Table 5.4.3

The influence of irrigation and mineral nutrition regimes on the accumulation of nitrates in 1 kg of corn mass

#	Fifth option	Nitrates, mg
Soil R_{zR} 60-55 %		
1	No fertilizer (co-ntrol)	175,1
2	$N_{90}P_{75}K_{60}$	260,8
3	$N_{120}P_{125}K_{90}$	395,4
4	$N_{200}P_{175}K_{120}$	402,1
Soil R_{zR} 70-65 %		
1	No fertilizer (co-ntrol)	120,0
2	$N_{90}P_{75}K_{60}$	188,0
3	$N_{120}P_{125}K_{90}$	209,3
4	$N_{200}P_{175}K_{120}$	281,4

The analysis presented in the table shows that nitrates accumulate in the smallest amount in the no-fertilizer variants, however, in the last two cases of using increasing fertilizer rates in low humidity conditions($N_{120}P_{125}K_{90}$^და $N_{200}P_{175}K_{120}$), the accumulation of nitrates in the green mass gave undesirable results and was within the limits of the maximum permissible concentration.

In the case of improving the irrigation regime, this indicator decreased to a harmless characteristic, which indicates the great importance of irrigation.

5.5. Chemical composition of the predicted corn crop, taking into account agronomic conditions and soil moisture

The results of the studies show that the impact of different types of fertilizers on the yield during irrigation, against the background of a fairly high level of biological nitrogen left by the predecessors, is directly related and depends on the type and combination of nutrients. Despite the insignificant changes in the yield depending on the variants, a significant effectiveness of this factor on the quality indicators of the resulting products is noted.

Many years of research by the Food Production Laboratory of the Department of Food Production and Animal Husbandry of the Georgian Agrarian University have established that the leading role in obtaining high-energy feed from corn grain and straw belongs to moderate plant nutrition with nitrogen. It is likely that mineral nitrogen is significantly washed out by irrigation and disappears, while the role of biological nitrogen in the formation of high-quality feed under such conditions inevitably increases.

It should be noted that the nutritional value of the products obtained by the food crop rotation was quite high in both the no-fertilizer and fertilizer variants. Table 5.5.1 presents the nutritional value of corn grain and straw, conducted only in the control and yield-enhancing variants #1 and #8, since the average data for determining the energy value of `Ajameti Tetri' are given sufficiently well according to these indicators.

Table 5.5.1

Composition and content of grain and straw in energy food units per kg of natural feed

Option number	Dry matter, g/kg	Raw food ingredients					Mineral substances		Digestibility					Energy Concentration (E.C.E)	Digestible crude protein, g/kg	Content		
		Protein	Fat	Cellular	Nitrogen-free extractive matter (UEN)	Ash	P	h	Energy	Crude protein	Crude fat	Raw cellulose	itrogen-free extractive matter (UN)			m.e.m.	(E.C.E)1 kg	Crude protein
1	2	3	4	5	6	7	8	9	10	11	12	13	14	15	16	17	18	19
Corn kernels																		
1	880	106	45	56	773	20	3,2	0,4	83	71	84	27	94	765	75	98	673	66
8	880	107	47	27	804	17	3,2	0,4	86	73	85	68	93	788	78	98	695	69
Corn straw																		
1	875	75	15	335	492	8,3	2,6	7,0	54	40	60	61	57	439	30	68	384	26
8	875	77	18	836	489	8,0	2,6	7,0	55	41	61	60	55	441	12	69	387	28

The biochemical analysis and calculation of nutrients using regression equations showed a high grain yield in the control, non-fertilized variant, where the difference between the energy food unit and digestible crude protein was only 1.03 and 1.05

percent compared to the eighth variant. Similar data were obtained in the case of straw, where the difference in food yield as a result of recalculation was 1.01 and 1.08 percent, respectively.

This can be explained by the fact that here the plant was provided with nitrogen, phosphorus and potassium, due to the mobilization of soil reserves of these basic nutritional elements.

In our experiments, a relative decrease in cell volume in corn grain and its increase in straw were noted, which is explained by the limiting factor of plant productivity. In addition, the positive effect of providing plants with basic nutrients was reflected in the fat content in grain and straw, which in the best variant, although insignificantly, still increased by 1.2%.

As for the content of crude nutrients, as expected, in the control variant, the ash content in both grain and straw increased.

According to the processing, in the calculated percentages, an increase in energy content, crude protein, crude fat and a decrease in crude fiber were observed in corn grain and straw. Energy concentration also increased significantly in the eighth variant, which indicates the development of the plant under similar ecological conditions.

5.6. Predictable corn crop yield

The energy content of food is the property of food to provide the livestock organism with the necessary substances, nutrients and energy.

Evaluation of the energy content of food and standardization of livestock feeding in terms of pure energy of fat production (OCT/BKC 663 food units, introduced in 1933) is very difficult, since it is based on the determination of the chemical composition by full zootechnical analysis and subjective selection of digestibility coefficients. The chemical composition is given in the reference book approximately.

At the modern stage, indirect methods for determining the nutritional value of feed have been developed and are used in production in many countries of the world, which are based on determining the correlation between some nutritional substances or their groups and the actual nutritional value of feed.

High productivity of livestock can be achieved if feed producers produce corn in feed grain rotation to obtain high-quality feed in large quantities, which in turn requires accurate and complex determination of the nutritional value of feed made from corn and its quality with a high level of organization of testing.

The nutritional value of food, expressed in food units or metabolic energy, is determined by the presence of the main nutritional elements in it, namely, the content of protein, fat, fiber and nitrogen-free extractive substances (NFES) taking into account their digestibility (A. Korakhashvili [12]). The content of these substances and their interdependence are determined by many factors. First of all, the vegetation phases of the agricultural crops taken as food, the agricultural techniques used, the doses of fertilizers applied, the technology of food production and storage, their preliminary preparation for livestock. The nutritional value and digestibility of various types of food depend on this.

It is impossible to assess the nutritional value of feed by any one indicator, since it cannot reflect the needs of the livestock organism for it. In recent years, a complex assessment of nutritional value has been used, taking into account the energy nutritional value of feed, dry matter, protein and its amino acid composition (amino acid composition), fat and unsaturated fatty acids, sugars, starch, fiber, macro and microelements, vitamins and other nutritional elements that are necessary for maintaining livestock health and obtaining high-quality livestock products. The nutritional value, or energy food unit (Table 5.1.1), in the control variant of corn grain is 673 kJ per kg, and in the best variant it is 695 kJ. In the control variant, the straw contains 384 kJ, and in the best variant – 387 kJ. Dry matter in grain is 880, in straw – 875, protein content in grain in the best variant is 107, in straw – 77, fat in grain – 47 g, in straw – 18 g, cellulose in grain – 27 g, in straw – 836, nitrogen-free extractive matter in corn grain, in the best variant – 804, in straw – 489.

Mineral substances play a very important role as a building material for the animal body (especially for bones and teeth) and as regulators of physiological processes in the body. They help maintain the optimal state of colloidal systems of cells and tissues. They ensure the penetration of nutrients and metabolic products into the cells, participate in the vital activity of the nervous and muscular systems, and determine the normal functioning of enzyme systems, some hormones, and vitamins.

Vitamins promote the absorption of nutrients, normal metabolism, improve the health of livestock, stimulate intensive weight gain in adolescents, increase productivity, and enhance the functions of self-reproduction of livestock.

Carbohydrates are the most common organic compounds in living nature: they account for more than 2/3 of the organic matter of plants, while the content of carbohydrates in the human body is much lower. Their main purpose is to produce energy during the breakdown of molecules, which is necessary for the life of livestock. In zootechnical analysis, two groups of substances are distinguished: cellulose and nitrogen-free extractive substances. In the diet of young ruminants, cellulose is considered a necessary component for the normal digestion of food. Since ancient times, plant foods contain a large amount of starch and sugar, which occupy the most important place in the nutrition of livestock. During the analysis of the composition of feed, crude fat is determined - a substance that is released during the processing of feed. It is included in livestock feed in the form of vegetable oils and fat-like substances. The quality of edible fat affects the quality of fat accumulated in the body of livestock.

Such progressive industrial technologies for assessing the nutritional value of food ensure the production and preparation of food that is uniform in physical and technological properties and balanced in nutritional substances, which will allow us to more fully utilize the automation and mechanization of farms, eliminate manual labor and rationally use modern technical means, and increase productivity and efficiency.

Chapter VI.
Economic Efficiency of Predictable Corn Crop

6.1 Justification of the Economic Efficiency of Predictable Corn Crop

We conducted an economic evaluation of corn growing methods on the example of field trials. Mainly according to economic indicators, which were based on corn grain and straw yields. We calculated the output of energy food units obtained per hectare by constructing a multi-year average of the total cost of the products obtained, since we adjusted the cost of cultivation annually due to high inflation during the period of conducting the trials.

To truly assess the effect of all trial options, we calculated all the costs associated with the use of the methods under study and brought grain corn.

The calculation of the methods to be studied was carried out on the basis of the methodology adopted for AMI, according to the technological maps of the farm, the reference book of mineral fertilizers ``Tariffs for growing corn", ``Retail prices for minerals, organic and biofertilizers", as well as ``Practical manual on the development of intensive technology for growing grain corn".

When analyzing the economic efficiency of the methods to be studied in the given crop rotation ring, it should be noted that the greatest net income from one hectare was obtained with the complex use of mineral fertilizers at 70-65% of the soil's marginal capacity in options 7 and 8. In these options, the lowest cost per unit of energy input was observed, i.e. 1.97 GEL less than in the control.

In terms of efficiency, the most profitable option turned out to be the high-yielding option – 151%. The profitability of the control options gradually decreased over the years. In the best option in terms of income, this indicator was almost stable. Therefore, we can conclude that at 70-65% of the marginal soil carrying capacity, when we applied mineral fertilizers at a dose of N120P125K90.^The corn grown in the 4th and 8th options is much more favorable not only from an ecological but also from an economic point of view.

Thus, the analysis of the economic efficiency of the studied methods and the marginal soil carrying capacity and fertilizer use showed that the cultivation of corn sharply increases the main economic indicator – profitability. The best economic results in terms of conditional net income are obtained in options 4, 7 and 8. In these options, the largest grain and straw yields are obtained at the lowest cost per unit of digestible crude protein and energy reserves. The control option is inferior to these options in terms of economic indicators, which proves the feasibility of alternative field food production. Based on the results obtained, the following conclusions can be drawn:

1. As a result of processing multi-year soil-climatic data, the existence of a linear correlation between the increasing sums of active temperatures and the corresponding radiation balances of agricultural crops, including corn, using modern methods of programming;

2. The complex coefficient of water supply was used to assess the water supply of the region, the absolute value of which expresses the tension caused by the lack or abundance of precipitation and, compared to the existing criteria, more fully reflects both the quantitative and qualitative aspects of the phenomenon;

3. Based on atmospheric precipitation, air temperature and other climatic and meteorological data, a new classification of water supply years is proposed based on the complex coefficient of water supply, which ensures the programmed yield of corn in the current climatic conditions at an average of 7.9 t/ha;

4. The distribution of the soil's maximum water holding capacity requires a differentiated determination of the moisture content of the active layer of the soil at different stages of corn growth and development and, accordingly, the specification of irrigation norms and regime, which should be maintained throughout the entire vegetation period;

5. Under both regimes of soil moisture before irrigation obtained in the field trial, the occurrence of phenological phases changes with increasing fertilizer rates and is directly related to the degree of moisture content, which has a significant impact on the main agronomic goal - corn yield.

6. When determining the degree of drainage, it is necessary to determine the density of water in the pores and, accordingly, aeration, which is one of the important factors for obtaining a programmed harvest, according to the percentage content of the maximum molecular water holding capacity of the soil;

7. Under high agrotechnical conditions, the combined effect of irrigation and fertilizing results in a much higher corn yield than when these factors act separately. In the first case, 2.85 t/ha, the increase is almost twice as high as the marginal yield of 6.10 t/ha in the second case. The effect of these two interactions increases both in connection with the fertilizing norms and the high pre-irrigation moisture content;

8. The high yield of corn grain and straw compared to the control option clearly indicates the great economic effect of programming, the best

9. The obtained corn crop, along with its abundance, is characterized by a good chemical composition and high energy content, which on average provides 7758.2

energy feed units for cattle (ESM) for the best option;

10. In order to obtain a programmed high corn harvest in the Mukhran-Saguramo zone, with the correct combination of existing soil-climatic conditions and the implemented agricultural techniques, we consider it appropriate to accurately take into account the main parameters of our research for each corn variety, which ensures the receipt of 7.90 tons of grain and 6.10 tons of straw per hectare in the case of `Ajameti Tetri`;

11. Implementation of recommended agrotechnical measures, along with increasing corn crop yields, will minimize the decrease in soil fertility due to excessive water. By eliminating ballast and other undesirable factors remaining in the soil as a result of the use of high doses of mineral fertilizers associated with irrigation, which will significantly reduce the "stress" on the ecological conditions of the environment.

References

1. Abashidze L. (1987) – Extraction of nutrients from the soil for obtaining a programmed harvest of grain corn in the irrigated areas of Shida Kartli. //Scientific works of the State Agricultural University of Georgia, Tbilisi, p. 351.
2. Abashidze L. (1990) – Dependence of the chemical composition of the programmed harvest of grain corn on the irrigation regime and doses of mineral fertilizers. //Scientific works of the State Agricultural University of Georgia, Tbilisi, p. 105.
3. Badrishvili G. (1981) – Crop production. Tbilisi.
4. Geliashvili P. (1985) – Fundamentals of land reclamation and irrigation practices. Tbilisi.
5. Goglidze E. (1961) – Methodical instructions on the use of agro-meteorological materials in socio-economic production. Sak. SSR Publishing House, Tbilisi. ^
6. Guntaishvili L. (1979) – Materials for the study of some physical and physico-mechanical properties of old alluvial soils of Kartli. Tbilisi.
7. Elizbarashvili E., Chavchanidze Z. (1992) – Droughts, with dry and rainy periods in Georgia. Tbilisi, Plantation, 110 p.
8. Zardalishvili O. (1978) – Basic principles of agriculture and the use of fertilizers. Tbilisi, "Soviet Georgia", 90 p.
9. Zardalishvili O., Nakaidze I. (1992) – Production and use of organic fertilizers. Tbilisi, "Georgia", 132 p.
10. Zardalishvili O., Urushadze T. (1992) – Fertilizer use and environment, Tbilisi, "Georgia", 132 p.
11. Kelenjeridze K. (1957) – Brief agroclimatic characteristics of corn growing regions of the Georgian SSR. // Proceedings of the Georgian SSR, vol. XI, Tbilisi, 43-62 p.
12. Korakhashvili A. (1991) – Modern methods of food quality assessment. Tbilisi.
13. Korakhashvili A. (1992) – Food production with programming elements. Tbilisi.
14. Kobaladze K. (1985) – The influence of irrigation on corn yield in the conditions of the Mukhrani Valley. // Proceedings of the Georgian Academy of Sciences, Tbilisi.
15. Kordzakhia M., Napetvaridze E. – General climatic characteristics of the territory of Georgia. /Book: Agricultural production of the Georgian SSR, specialization and agricultural management systems, Tbilisi, `Soviet Georgia', 28-53 pp.

16. Kordzakhia M. (1961) – Climate of Georgia, Tbilisi, Publishing House of the Georgian SSR Academy of Sciences, Tbilisi, 243-246 pp.
17. Kotaria A. (1972) – General meteorology and climatology. Part I, Tbilisi, TSU Publishing House, , 373 p.
18. Machavariani J. (1980) – Soils of Shida Kartli. TBI-LIS, 920 p.
19. Naskidashvili P., Liparteliani O., Jinjikhadze Z., Kapatadze G., Saatashvili I. (1993) – Results of selective use of indigenous varieties of corn in Georgia. //The Academy of Agricultural Sciences of the Republic of Georgia, Gene pool of agricultural plants and animals, its protection and use (Materials of the scientific-practical conference), Tbilisi.
20. Saatashvili I. (1995) – Corn of Georgia Obtaining self-pollinated hybrids from aboriginal varietal populations and creating highly heterozygous hybrids based on them. Doctoral thesis, Tbilisi.
21. Saatashvili I. (1974) – Testing of different types of corn hybrids in different ecological conditions. // Proceedings of the Georgian State Agricultural University, Vol. #88, Tbilisi.
22. Sikharulidze M. (1975) – Selection and breeding of field crops. Tbilisi, `Education'.
23. Talakhadze G., Anjaparidze I., Lataria V., Kirvalidze R., Nakashidze N., Mindeli M. (1983) – Soils of Georgia. Tbilisi, `Education', 355 p.
24. Tughushi G. (1986) – Rules for irrigation of agricultural crops and ways to improve them. Tbilisi, `Soviet Georgia~.
25. Tughushi G., Abashidze L. (1990) – Determination of water requirements for the planned harvest of grain corn in terms of soil and climatic factors. //Scientific works of the Georgian State Agricultural University, Tbilisi.
26. Tughushi G., Abashidze L. (1996) – Regularities of the release of nutrients by the corn crop under different irrigation regimes and soil fertilization norms with mineral fertilizers and their analytical generalization. //Scientific works of the Georgian State Agricultural University, Tbilisi, p. 20.
27. Tughushi G., Abashidze L. (1986) – Determination of the irrigation regime of corn for obtaining a programmed harvest in the conditions of the Mukhran-Saguramo field. //Proceedings of the Institute of Agricultural Sciences, Tbilisi.
28. Tughushi G., Abashidze L. (1990) – Programming of grain corn yield under irrigation conditions taking into account meteorological influence. //Proceedings of the International Conference, Kestney, Hungary.
29. Tughushi G. (1990) – Fundamentals of statistical analysis of the irrigation regime

of agricultural crops. //Proceedings of the Institute of Agricultural Sciences, Tbilisi.
30. Tughushi G. (1986) – Agricultural land reclamation, design of irrigation systems, part 1, TBI-LIS.
31. Tughushi G. (1993) – Determination of the intensity of soil water erosion and assessment of the potential erosion risk. //SAU-s Shrombebi, pp. 162-167.
32. Chkhenkeli I. (1970) – Irrigation technology. Tbilisi.
33. Tsagurishvili G., Keshelashvili E., Mtvarelishvili Sh., Peradze I., Manjavidze M. (1990) – Agriculture. Tbilisi, `Public Education', 405 p.
34. Tsutsunashvili O. (1985) – Agricultural melioration practicum. Tbilisi.
35. Japaridze A. (1970) – Corn culture in Georgia. Tbilisi,
36.Chanishvili Sh. (1979) – Crop programming. //Journal `Agriculture of Georgia~, #4, pp. 13-14.
37. Chavchanidze Z., Yanakosidi M., Elizbarashvili E. (1994) ¬– Research of soil resources of Georgia. //I International Scientific Conference dedicated to the Pontic Greeks, Tbilisi.
38. Arveladze B.A. (1986) – Agrometeorological aspects of tea leaf crop formation. M., Гидрометиздат, стр. 12-35.
39. Бадина Г.В., Королев А.В., Королева Р.О. (1988) – Основы агроно¬мии. "Агропромиздат", Leningrad, , 444 стр.
40. Bakhtizin N.R. (1978). /В кн.: Теория и практика прог¬раммирования урожаев. М., ВАСХНИЛ, стр. 68-69.
41. Berbekov П.Л., Каюмов М.К. and others (1980). 77-83.
42. Bondarenko Н.Ф., Константинов А.Р. (1981) – Пути опти¬миза¬ции режимов орошения. //"Гидротехника и мелиорация", №11, стр. 64-67.
43. Bondarenko N.F. (1981) – Basics of cropping on drained lands. //"Гидротехника и мелиорация", №11, стр. 64-67.
44. Udagovskii A.I., Lozovskaya E.A. (1976) – Theory of summarization and its applications. //"Водные ресурсы", №2, стр. 34-56.
45. Vorobyev E.C., Voronkova Ф.В., Titov B.C, Pronin B.A. (1981). Leningrad, "Колос", 101 str.
46. Voronin N.G. (1989) – Орошаемое земеледелие. М., "Агро¬промиз¬дат", 335 стр.
47. Demchenko Е.И. (1988). //"Мелиорация и водное хозяйст¬во", Експресс-информация, Орошение и раститель¬ные системы, М., стр. 15-23.
48. Димитриева З.А. (1980). /В кн.: Ос¬нов¬ные фактори программирования,

Riga, str.42-46.

49. Dovnar V.S. (1970). /В кн.: Важнейшие проблемы фото¬син¬теза в растеноводстве, М., "Колос", стр. 198-316.
50. Иванов А.Ф., Naumov H.A. (1978) – Productivity of different hybrids of corn. //Тр. Волгоградского СХИ, вып. 65, str. 146-150.
51. Иванов А.Ф. (1979). //Тр. Волгоградского СХИ, , вып. 71, str. 19-24.
52. Kayumov M.K. (1975) – Урожай по заданной программе. //"Кадры сельского хозяйства", М., №1, стр. 95.
53. Kayumov M.K. (1976) – Experience in obtaining planned harvests. /В кн.: "Программирование урожаев сельскохо¬зяй¬ст¬венных кул¬тур", Chişinău, str.48-56.
54. Kayumov M.K., Verbitskaya M.M. (1978) – Programming of grain crops, M.
55. Klevtsov M.N. (1980). //Материалы Всесо¬юз¬ной конф., Нальчик, Ельбрус, стр. 88-94.
56. Константинов А.Р., Strünnikov Э.А. (1986). //"Гидро¬тех¬ника и мелиорация", №1, стр. 19-28, str. 33-42, No. 3, p. 37-44.
57. Кружилина Ж.В. (1987) – Growing corn on grain in intensive technology. Volgograd.
58. Kuznetsov H.И., Опенгейдер И.В. (1971) – Effect of boron on yield and quality of corn. //"Агрохимия", №2, стр. 142-144.
59. Kulakovskaya T.N. (1975) – Programming of high-level agricultural crops. Methodical recommendations. Минск, 42 str.
60. Kulakovskaya T.N. (1978). Минск, "Урожай", 272 стр.
61. Кулаковская Т.Н., Шаталова Р.В. (1980). /В кн.: Ос¬нов¬ные фактори программирования, Рига, стр. 76-81.
62. Листопад Г.Е. and others (1976). Volgograd, 34 str.
63. Lьgov Г.В. (1973). //"Гидро¬техника и мелиорация", №4, стр. 50-53.
64. Maslov B.S. (1984) – Agricultural reclamation, M., "Колос", 510 стр.
65. Matveev I.T. (1987) – General meteorology. M., "Гидроме¬тео¬из¬дат", 701 стр.
66. Meladze G.G. (1991) – Ecological factors and production of agricultural culture. L., 167 str.
67. Михайлов В.В., Михайлова Т.Д. (1975). //"Гидро¬тех¬ника и мелио¬рация," №6, стр. 87-89.
68. Muhamedov E.M. (1980). /В кн.: Основ¬ные фактори программирования, Рига, стр. 63-69.

69. Нерпин С.В., Чудновский А.П. (1975).
70. Obraztsov L.S. (1980) – System application in agriculture. М., "Агропромиздат", 302 стр.
71. Pavlova N.D. (1984) – Практикум по агрометеорологии, Л., "Гидроме¬тео¬из¬дат", 184 стр.
72. Papalashvili G.M (1975) – Одностебельно-много¬поч¬ко-вость куку¬рузы. //"Corn", No. 1.
73. Plyusnin I.I., Golovanov A.I. (1983) – Мелиоративное поч¬воведе¬ние. М., "Колос", 318 стр.
74. Polozova L.G. (1981) – Probability of droughts in the main agricultural products of the northern hemisphere. //Тр. Гл. Geophys. observ., №458, 101-105 str.
75. Popova T.P., Malysheva P.A., Emelyanova I.M. (1983). /В сб.: Методы полевых иследования по отвечения мелио¬ра¬ции, Л., стр. 256.
76. Радов А.С, Stolipin E.H. (1978) – Удобрение в рошаемом земледелии., М., "Наука", 224 стр.
77. Ross Yu.K. (1977) – Radiation regime and architectonic vegetation cover.
78. Savitsky M.S. (1978). //"Химизация социалистического земледе¬лия", №7, 45-47 стр.
79. Синякова Л.А., Васко В.Т., Зайцев В.Я., Ганускевич Ф.Ф. (1987). Л., "Агропромиздат", 223 стр.
80. Сиротенко С.Д. (1981) – Mathematical modeling of water-thermal regime and productivity of agroecosystems. Л., "Гидроме¬тео¬из¬дат", 167 стр.
81. Сурин В.А., Носенко В.Ф. (1981) – Mechanization and automatic irrigation of agricultural crops. М., "Колос", стр. 151-220.
82. Tooming Kh.G. (1978). str. 10-17.
83. Ulanova E.C. (1975) – Agrometeorological conditions and yield of wheat. Л., "Гидроме¬тео¬из¬дат", 302 стр.
84. Fortinskaya G.S. (1975) – Effect of interaction of irrigation and fertilization. //"Гидротехника и мелиорация", №4, стр. 109-113.
85. Циприс Д.В., Шевелев Я.З., Ревут В.И. (1978). Л., "Колос", 113 стр.
86. Tsertsvadze Sh.I., Stolipin N.P. (1959) – Agroclimatic characteristics of corn cultivation in Transcaucasia. //Тр. Тбил-НГМИ, вып.4, стр. 172-173.
87. Чирков Ю.И. (1972).
88. Shatilov I.S. (1973) – Принципы программирования уро¬жай¬нос¬ти, //Вест¬ник слежастияных наук, №3, стр. 8-14.
89. Shatilov I.S. (1975) – Principles of programming уро¬жай¬нос¬ти. /В кн.:

Программирование урожая сельскохозяйственных култур, М., "Колос", стр. 17.
90. Shatilov I.S. (1979) - Maximal accumulation of solar energy by cultural plants - the most important task of modern agriculture. /В кн.: Вопросы интенси¬фи¬ка¬ции земеледелия, Yoshkar-Ola, стр. 72-82.
91. Shatilov I.S. (1976).
92. Shatilov I.S, Kayumov M.K. (1977) – Программирование урожаев в колхожах и совхозах //"Земледелие", №1, стр. 54-58.
93. Shatilov I.S, Kayumov M.K. (1978). М., "Колос", 332 стр.
94. Shatilov I.S, Kayumov M.K. (1979) – Программирования урожаев полевых култур. Methodical recommendations, М., ВАСХНИЛ, 88 стр.
95. Elizbarashvili Э.Ш., Chavchanidze З.Б., Yanakosidi M.C. (1994).
96. Ms. Qing F.D. (1980) – Climate variability and local temperature regions in food climate interactions, Proc. Int. Pop. Berlin, Dec. pp. 9-12
97. Klein C.H. (1981) – Application of statistical. heather broadcasting to different time scales. //Publ. Ch. Meleoland Godkn, Chln, №253, pp. 159-162.
98. Stern D. (1982) – Analyzing daily rainfall increments to give agronomically useful results. //Ekhp. Agr., 18, №3, pp. 237 253.
99. Martin P.N. Gent and Herbert I. Enock (1983) – Temperature Dependence of Vegetative Growth and Dark Respiration. A Mathematical Model Plan Physiol., 71, pp. 562-567.
100. Holt D.A. (1975) – At All Environmental Physiological Modeling and Simulation of Alfalfa Growth. //Conceptual Development of SHIMED, Prod., Agric. Ekhp. Sht. Ges. Bull, №907, p. 26.
101. Bayer Ch. (1981) – Chrop–Cheather analksis model. //"Appl. Ghemote Shens. Agr. Prod. Forestast Ilect. Chourse", Zhaint. Ges. Cent., pp. 105-118.
102. Kobaladze K.A. (1990) – The mineral nutrition and irrigation regime of forage direction triticale in the conditions of Mukhrani. //XXXII Georg., Tudom. Kekzthelk., pp. 210-214.
103. Eberhard Sh. – Linear Programming Model Allowing for Complex Changes and Mutual Dependence of Coefficients in the Roots of a Plant.
104. Gara Sh. (1990) – Program Packaging to Support the Roots of Our Roots Plants. //XXXII Georg.. Tudom. - Keqzthelk., pp. 50-53.
105. Hendrich E.M.T. (1990) – Linking Linear Programming to Simulation Models: Some Experiments for Gattle Husbandry, //XXXII Georg. Tudom. - Keqzthelk., pp. 251-295.

106. Gitschter K., Kersbaum C. (1990) – Modeling Nitrogen Regime of Churable Giles for Advisory Purposes. //XXXII George. Tudom. Keqzthelk., pp. 152-158.
107. Kozma E., Duncan D. and Others (1990) – Computer Aided Coordination System of Raw Material Production and Progress in the Channing Industries Using Meteorological and Phenological Information. //XXXII George. Tudom. - Keqzthelk., pp. 17-21.
108. Fekrhern A.M., Pausen G.H. (1981) – Development of a Heat Yield Production Model. //Agron. J., №2, pp. 277-282.
109. Geknoldz J.F., Thurnleq J.H.M. (1982) – Shot: Heat Partitioning Model. //Ann.B.,

Electronic Library

1.http://tsssss.kvirispalitra.ge/.../44886-qqarthuli-simindi-ualternativ.
2. tssss.sakpatenti.gov.ge/media/
3. tssss.opentech.org.ge/index.php?m=12&q=2012&art=6033
4. tssss.guryanetss.tsom/article/karmidamo-tshemi/836
5. https://gfa.org.ge/
6. tssss.guryanetss.tsom/article/karmidamo-tshemi/836~
7. tsssss.interpressnetss.ge/
8. https://forum.ge/index.
Publisher: Ind. Entrepreneur Gocha Dalakishvili;E
E-mail: gotcha.dalakishvili@kahoo.tsom

Olga Kharaishvili Candidate of Agricultural Sciences, Associate Professor (author of 183 scientific papers) in the field of agricultural land reclamation

FSC
www.fsc.org
MIX
Papier | Fördert gute Waldnutzung
FSC® C083411

Kolibri 360 GmbH
Ferdinand-Jühlke-Straße 7
99095 Erfurt, Deutschland
produktsicherheit@kolibri360.de

Druck:
CPI Druckdienstleistungen GmbH
im Auftrag der
Kolibri 360 GmbH
Ferdinand-Jühlke-Str. 7
99095 Erfurt